In Pursuit of
Technological Excellence

In Pursuit of
Technological Excellence

Engineering Leadership, Technological Change, and Economic Development

Ernst G. Frankel

PRAEGER

Westport, Connecticut
London

Library of Congress Cataloging-in-Publication Data

Frankel, Ernst G.
   In pursuit of technological excellence : engineering leadership,
technological change, and economic development / Ernst G. Frankel.
     p.   cm.
   Includes bibliographical references and index.
   ISBN 0–275–94476–X (alk. paper)
   1. Engineers. 2. Engineering—Management. 3. Engineering
economy. I. Title.
TA157.F72 1993
620′.0023—dc20   92–23457

British Library Cataloguing in Publication Data is available.

Library of Congress Catalog Card Number: 92–23457
ISBN: 0–275–94476–X

First published in 1993

Praeger Publishers, 88 Post Road West, Westport, CT 06881
An imprint of Greenwood Publishing Group, Inc.

Printed in the United States of America

The paper used in this book complies with the
Permanent Paper Standard issued by the National
Information Standards Organization (Z39.48–1984).

10 9 8 7 6 5 4 3 2 1

Contents

Illustrations

TABLES

Preface

I do not know what ever enticed me to write this book. Was it a desire to explain why I chose to become an engineer even though there were many easier ways to make a living and certainly better ways to make a comfortable living? Did I have to expose my soul to convince myself of the beauty of engineering and the excitement of doing something real and meaningful? Something that serves? Something that performs a function? Or did I need an excuse to show my contempt for the multitude of people who call themselves professionals while contributing nothing real, of value, or useful to society? Engineers seldom are perceived or treated as professionals by the public, although they are largely responsible for the advances in our standard of living. Maybe my desire is to foster the recognition of engineers and their contribution to humankind and our way of life.

Most of what we do, use, or enjoy requires engineering. Today, even basic food production, the construction of dwellings, and transportation are heavily dependent on engineered technology.

While most forget to credit technology and thereby engineers for the many advances we enjoy, too many are quick to blame engineers for the pollution of the environment and the impact of technology on our way of life, more often forgetting that the choice of adopting or introducing technology was seldom made or demanded by engineers. In fact, engineers usually warn of the dangers of technology only to be overridden by more powerful laypersons who take advantage of what technology has to offer.

We are at a crossroads now, and the future points to many risks and opportunities: risks that technology will be misused and drag the world and its environment down and opportunities for a better and brighter life for humankind in an ecologically clean world. Engineers must play an increasingly important role in the future development of the world if we are to make better choices than in the past, because only they can judge technology, its capabilities, and its impacts. The future well-being of humankind will become more dependent on the quality of the technological choices and the use of technology as time goes on. Without a change in the role engineers play in formulating policy and in making decisions, technology likely will be misused and the pursuit of technological excellence in the service of humankind will become elusive. We need engineers to guide us into a better technology-led world, and we need engineers able to provide such leadership. The satisfaction of these needs will require a drastic change in the self-perception, education, and personality—as well as the role—of engineers.

We generally recognize that technology has become the principal driving force of economic development and technological change, but the single most important challenge for national leadership and the most difficult decision problem for management in general—the role and function engineers play in technology development and technology management—remains undefined and ignored.

Most people, including government and industrial decision makers, have only a very vague idea of how technology is developed and used. Most technological developments are taken for granted, with little effort exerted to understand the underlying processes and factors involved. Most people are ignorant of the physical, technical, and economic principles involved, as well as of how and why the technology was developed in the first place. The basic creativity, purpose, and objectives that advance technological development are nearly always forgotten in the drive of a technology-user society that is interested only in whether technology can advance the users' or decision makers' objectives and does not care why something works, what makes it work, or how the technology that makes it work was developed.

It is suggested that an increasingly technology-dependent society must reevaluate and reestablish the role, function, and education of engineers if it is to maintain technological and economic leadership. While many in government and management recognize the need for and contribution of engineers, the majority of such leaders and the public at large not only have little conception of what engineers do but do not even understand or care about the role of engineers in the development and use of technology.

This is a book about engineering, about its purpose and content, about the way engineering is done, and about how technological excellence is achieved. It represents my views as someone who has practiced engineering for over 40 years and who has attempted, often unsuccessfully, to achieve the basic objectives of engineering as I see them. These objectives are to develop technology, improve it, put it to use, and teach how and when it is to be used, including the constraints inherent in the technology. Engineering includes technological decision making and the effective management of technology.

I do not perceive engineering to end with the research and development of technology or the design of some technology. Engineers are technology developers, designers, makers, and trainers. Their principal function is to develop effective uses of science or scientific principles in terms of some potentially useful technology; to design the technology to develop a method of building, manufacturing, or otherwise implementing it; and to establish effective uses for it. Engineers are the ultimate integrators, and their principal calling is the development and integration of the various activities required to bring a technological opportunity, based on some discovery, to ultimate use.

Engineers who consider their role more restrictive usually end up as narrow engineering scientists, designers, industrial or operating engineers, with very limited roles and, most important, practically no decision- or policy-making powers. They become hired hands or confined investigators. Either they are given tasks by others usually not trained in engineering, such as managers or administrators, or they work on narrow problems defined by them or others. They play no role in if, how, when, where, and by whom the results of their investigation or their infant technology are to be used.

Engineers in the Western world, particularly the United States, have become narrow and specialized. Most see their function as performing some particular role in research, investigation, design, manufacturing, or operating technology. This function seldom is defined by technical persons. Engineering as a profession, as a result, is difficult to define, and some may argue that it really is not a profession. Its scope varies from highly sophisticated and often abstract scientific investigation—technology development and engineering design—to technical management, repair, and operation. The public's perception of engineers is even more vague. Auto mechanics, boiler operators, designers of bridges and other infrastructure, as well as applications engineers are considered to be engineers.

Few today would call some of our greatest inventors—such as Alexander Graham Bell, Henry Ford, and James Watts—engineers, although most were engineers by choice. These are curious developments, particularly in a country such as the United States which throughout its history benefited from the inventiveness of its engineers and which surpassed the industrial powers of Europe in productivity by sheer technological originality and the courage and creativity of its engineers. It is probably no exaggeration to say that the United States owes its greatness to the inventiveness of its engineers and their ability not only to invent technology but to show how it could be implemented and used. Our engineering forefathers were not narrow engineering scientists or designers. They were technology developers and integrators who changed our way of life, the world around us, and our role as a nation.

This condition is no longer so. We still make scientific and technological discoveries, but we are more and more unable to translate them into actual and useful applications in a timely or effective manner. Our engineers and engineering scientists have become narrow specialists who usually do not recognize or care about technological opportunities. They consider their role as narrow contributors to knowledge of technology without any responsibility for its broader development or use.

This situation is the result of the focus of our narrow, often highly specialized engineering education, the structure of our industry and society, the perception of engineers by society, and the self-perception of engineers. It is caused by our decision- and policy-making hierarchy in government, industry, academic institutions, and society in which engineers seldom assume a leading policy- or decision-making role. But most important, it is the result of the way engineers live, work, communicate, interact, organize themselves, and respect themselves that affects their contribution to society. While technology drives nearly everything, the role of engineers is actually less important in decision-making terms today than it was at earlier times, when technology was barely recognized as an important contributor to society.

Unfortunately, the gap in knowledge between technology developers and those deciding on its use—including the ultimate users—continues to grow. Technology is becoming increasingly complex and pervasive, with ever-larger impacts on society, our way of life, and the environment.

Engineers will have to participate in or even lead the policy- and decision-making processes, particularly those affecting or affected by technology. In a modern society, engineers must assume more responsibility and face the issues created by technology because only they truly understand the capability of technology, its potentials, dangers, and opportunities for development.

Seldom do engineers take the responsibility for technological decisions. While not taking responsibility may be the result of a lack of opportunity to do so, because so few engineers are in senior management or government positions, it more likely is due to the basic unwillingness of engineers to take responsibility and make difficult, often risky, choices. Engineers are trained to be risk averse and to make decisions only when outcomes can be predicted with reasonable certainty.

Engineering education is at least partially to blame, as it does not encourage engineers to take risks and does not train them in management or decision making. Engineers similarly are ill-equipped as communicators and—as a result—negotiators.

In the past, engineers were creators who developed new concepts or devices and used their mental and physical skill to come up with solutions to problems or new technologies. They thought in terms of products, processes, or services. By contrast, engineers in the West today have degenerated into operating engineers, design engineers, or engineering scientists. Few engineers work independently and most serve decision makers in other disciplines who may have little, if any, technical knowledge. In other words, engineers have become solvers of problems set or chosen by others who often have little knowledge or appreciation of technology—except for its potential users.

For engineers to play an essential role in society and assure effective use of technology for the good of society will require a radical change in engineering education and the role of engineers in society. Engineering education must challenge engineers to uncover opportunities, recognize the limitations of the environment in which technology works, judge economic and social implications of technology use, and realistically integrate technological developments in the real world with all its frailties, uncertainties, and consequent risks.

Engineers will have to learn that they cannot cover up risks or lack of knowledge by safety factors but that risks must be confronted by rational analysis of the range and level of uncertainties. Engineers will need to acquire greater skills in communication, systems analysis, and management.

Education in scientific and engineering principles and theory does not have to be downgraded or diluted, but the teaching of these principles and theories must take place in the framework or context of the real and not ideal world.

In this book I attempt to evaluate engineering as a profession and engineers as individuals. Suggestions are offered on how the role and function of engineers could be moved into the mainstream of both short-term and strategic decision making in government, industry, and society and how their contributions can be emphasized. Most important, the link between technological excellence and engineering leadership in society is argued and a menu is proposed for a more effective, compassionate, environmentally conscious, and responsible technological society.

Many technological decisions are made today by leaders without any technical background as well as without the benefit of engineering advice or knowledge. This situation not only is hazardous in both the short and long run but also affects performance of industrial and economic development, as many policy decisions are based on ignorance of technological capability and impacts.

There obviously are many exceptions to the lack of technological excellence and engineering leadership presented here, yet it appears that if the advanced Western societies of Europe and North America are to maintain their technological excellence and economic progress—which, in the past, gave them their political and economic leadership—radical changes in the concept and role of engineering are required.

This book attempts to evaluate the role of engineering and the state of technology development in Western countries, particularly the United States, and offers a blueprint for the better use of valuable engineering and technological resources.

Acknowledgments

I am deeply indebted to two mentors who guided me over many years in the understanding of the role of technology in the solution of humankind's problems.

Dr. Frank Davidson, director of the Macro Engineering Research Group at MIT, showed me how technology could be used to address significant problems, while Dean Alfred Keil guided me in understanding the concept of the wise use of engineering.

While responsibility for the detailed matter in this book is certainly mine, credit for the basic idea of the role of engineering in advancing technology for the good of humankind and this earth rests with my mentors.

My lovely wife, Inna, suffered patiently the many weekends spent on this manuscript and contributed greatly by incisive comments on critical issues and by honing my understanding for and patience with human weaknesses.

Last, but not least, my gratitude to my long-term associate, Sheila McNary, who painstakingly drafted and edited this work.

In Pursuit of
Technological Excellence

1

Introduction

Technological change poses both threats and opportunities to society and its various institutions. The rate of technological innovation and resulting technological change has greatly accelerated in recent years and has become the dominant factor in economic and industrial development. Today, the management of technological change is probably among the most important issues confronting government, management, and individuals.

The management of technological change depends on the expected impact of technological change on society, its economy, its competitiveness, and its use of resources. It similarly impacts on defense, quality of life, and related issues affecting society. The rate of technological change has increased rapidly in recent years. The strategies used to introduce technological change and the management of the process of technological change often determine the success or failure not only of the resulting transition but also of the nation, industry, firm, or individual involved.

Managing technological change has been a haphazard process in many societies, few of which have developed plans for dealing effectively with such change. As a result, the degree of effectiveness of various management policies differs greatly among countries. For example, in the industrialized and newly industrialized nations of the Far East, policies are cooperative, streamlined, and responsive. But in many Western nations, the management of technological change is often performed in a disorganized, noncooperative manner and subject to litigious pressures.

Many reasons exist for the lack of effective planning for and management of technological change at the various levels of government and industry in different countries. Among them is the real and perceived risk of technological change. The risk may depend on uncertainties involved in the diffusion and acquisition process, on inadequate technology evaluation, on a lack of control of the rate of technological change and introduction, on uncertainties in timing, on competitiveness, on market share, on uncertainty in societal needs as well as its perception and requirements, on the societal level of technology, on a lack of government encouragement, and on regulation, among other factors. The more rapid the rate of technological change, the more resources that must be committed to the process of technological change. As a result, the risk is usually larger.

The institutional approach toward technological change in different countries is usually diverse. In general, technological change is recognized as an important element in the framework of industrial or economic development, but the evolution of technology is often quite different. Some countries attempt to use research to achieve technological advance; others—at least during formative years—import technology and adapt it to their requirements. The organization of and relations between labor and management similarly are quite diverse, as are the relations between industry and government. The cultural background and traditions of peoples often profoundly influence their acceptance of technological change.

Another important difference among countries is the development of decision-making hierarchies and organizational infrastructure. This development impacts not only on the effectiveness of the management of technological change but also on the long-term abilities of a nation or industry to maintain its economic and industrial growth.

Among the most important factors influencing the effectiveness of technological change is the role and function of engineers. In some societies, engineers attain senior positions in government and industry, but in many Western countries, they are relegated mostly to nondecision-making positions. Government leadership in both legislative and executive positions is more often assumed by lawyers, career bureaucrats, and politicians, while industrial management is largely the domain of professional managers, lawyers, financial experts, marketing specialists, and accountants. This makeup holds true even in government and industry leadership positions that are concerned largely with technological decisions, such as in the military and in fields like transportation and energy. Similarly, in industry, engineers usually are relegated to functional positions instead of senior line and strategic management positions.

This point is illustrated by the command structure of the U.S. military (civilian and uniform). The career path of engineering officers and civilians is severely constrained and limited to positions many levels below the top ranks in every branch of the services as well as in the service departments and the Department of Defense. The same applies to government departments in transportation, energy, and commerce, all of which today are concerned primarily with both the short-term and strategic management of technology. Engineers can only advance beyond their limited engineering career levels if they opt out of engineering into management, even though higher-level decisions involve mainly technological issues.

The low regard in which engineers are held also is apparent in industry, including high-technology industry in the United States and other Western countries. Here again, engineers are "tolerated" or used at lower levels in the hierarchy; they seldom are able to gain or retain positions in upper management even when they are the originators of the technology on which the industry relies or when they are the founders of the company. In many Far Eastern countries, by contrast, engineers attain some of the most senior positions in industry and government, and the management of technology is recognized to be the most important of management functions, with financial control, accounting, marketing, and so on, delegated to hired professionals in their respective disciplines.

As an engineer, I always had assumed that the important contribution of engineers to the advancement of society and human well-being was obvious to the public, particularly when technology permeates all aspects of everyday life. It came as a rude awakening to me, therefore, when I appeared before the high school class of my son to discuss the role of engineers and engineering only to find out that most of these 16-year-old teenagers not only were ignorant of what engineers do or what engineering is all about but perceived engineers as only operators and maintainers of technology. The exceptions were students whose parents were engineers.

The students thought that engineers drive locomotives; repair automobiles; operate boilers in apartment buildings; maintain highways and bridges; operate ships and aircraft; process equipment; and design structures, transport vehicles, and various processes. The view of the large majority was that engineers "make things work" and maintain things so that they work effectively. Only a small minority voiced the opinion that engineers are practical scientists trained to design things. No one felt that engineers should play a role in managing and running government departments or industrial firms concerned with technology or be involved in the development of technology.

I do not think they actually believed me when I told them that everything they used—from videocassette recorders (VCRs) and personal computers (PCs) to kitchen appliances and transport vehicles—had to be designed by an engineer, using advanced theories and principles of engineering design.

I did not convince the students that without engineers many of these technological developments would not be available for use. They perceived some isolated scientist generating a scientific breakthrough which then, as if by magic, appeared as a new useful technology a little while later. Most students thought, for example, that architects design buildings; only a very few knew that civil engineers were needed to develop the engineering design that permits the construction. Students knew much more about the role of lawyers, medical doctors, accountants, and entertainers who, in their opinion, were qualified professionals with very important roles in society and with well-defined responsibilities.

This perception of engineers and engineering in Western countries is much more widespread than I ever imagined, as I learned from subsequent discussions with friends and acquaintances, many of whom were themselves lawyers, doctors, accountants, or managers.

I was able to question senior government officials, and later industry managers, on the participation of engineering experts in strategic decision making affected by or affecting technology. Their replies were an incongruous stare, a noisy discussion, and an expression of disbelief that such a question could even be raised, as engineers obviously had little, if any, contribution to make to strategic management decision making. In this context, it should not be surprising that only a handful of engineers reach senior government and industry positions. In the United States, for example, less than 4 percent of the members of the House of Representatives and the Senate have an engineering background and an even smaller percentage ever practiced engineering. Yet nearly 33 percent of the budgetary decisions made by Congress usually deal with technology—including defense, energy, transportation, health care, and similar technology procurements. Industry in the United States and most other Western nations also does not have large numbers of engineers in upper management.

There is increasing evidence, however, that technical competence in upper management or among senior decision makers exerts a profound influence on the timeliness of technological decisions, as well as on the choice and the excellence of technological developments. In fact, there appears to be a direct correlation between the effective management of technology and the technological competence of decision makers.

Technological excellence cannot be achieved without the effective management of technology, particularly as the rate of technological development accelerates and as technology emerges ever more rapidly from the confines of narrow technological disciplines. We increasingly find in unrelated fields solutions to technological problems or opportunities for technology application. To manage new developments and assure excellence requires more than a superficial knowledge of technological capabilities and needs; it requires solid engineering education and experience. For engineers to serve as leaders and to assume the role of decision makers—not just the role of technology developers—will require a radical change in engineering education, engineering attitudes, and the role and function of engineers.

In this book, an attempt is made to study this issue and to consider its implications, both in the short and the long term. In particular, an attempt is made to consider if a change in the role, status, and education of engineers is desirable and, if so, how it could be achieved. The effects of such change on government and industry are projected, and both societal and economic impacts are evaluated.

The characteristics of engineers and the engineering profession are reviewed, and the way engineers solve problems and communicate is studied. It is curious to note, for example, that the engineers who design spacecraft, deep submergence vehicles, and intelligent systems are very conservative and risk averse. It will be interesting to consider why engineers who seek recognition, prestige, and status find that these rewards always elude them. Engineers are seldom in the role of the decision maker, and they usually lack the required communication and interpersonal skills to gain the rewards they seek. Yet they seldom make an effort to correct these deficiencies. They do, however, often assume responsibility or blame for technological or operational failures. Only by moving engineers into technological management positions in government and industry will engineers be able to emerge from technological isolation, make more timely and effective technological decisions, and regain or maintain technological leadership. The lessons learned from our past are described, and the requirements for sustainable industrial leadership and economic growth are established in a set of recommendations drawn widely from the experience of successful industrial enterprises and public agencies that have learned to cope effectively with the management of technological change.

While some countries, particularly in the Far East, have in the past built much of their industry on U.S. and European inventions, innovations, or techniques, they have long since passed the need of copying technology. Yet many in the United States and Europe still believe in the superiority of their own technological development processes. It well may be that many basic engineering and scientific breakthroughs still originate in the United States and Europe, but the process of technological innovation seems to be much less efficiently applied there than in the industrial countries of the Far East. The single most important factor for this discrepancy—accounting in large part for the large differences in the rate of economic growth between the older and newer industrialized countries—is the different role engineers play in their respective country's governments and industries.

The United States, as well as most Western industrialized nations, is technology based. Technological innovation and technology adoption have been the largest factors in the growth of the economies of these nations. Similarly, declines in the rate of economic growth in the United States, Britain, and certain other industrialized nations appear to be largely the result of a decline in the rate of technological innovation. Although the number of U.S. patents filed per year has remained reasonably constant, nearly half of all patents filed are now granted to foreigners. In addition, over 80 percent of the remaining applications are made by major U.S. companies, which traditionally lagged behind private inventors in the rate of innovations of patents filed. In other words, while the inventiveness of individual Americans may not be stifled, the number seeking patent protection has seriously declined. This may be due in part to the smaller number of engineers working as individuals.

The competitive position of the industrialized nations in Western Europe and, more recently, in the United States is deteriorating, notwithstanding increasing expenditures for research and development and for education and capital assets. A large percentage of inventions and discoveries made in these countries does not lead to new processes or products for many years. New processes and products often are introduced only after competing industrialized countries have been able to apply and improve upon the new technology. The impediments to a quick introduction of new processes and products include the slow progress in innovation and implementation, a lack of or insufficient support for technology development, the conflicting interests of the underwriters who often prefer to hide than to develop a new technology, and the perceived lack of interest by the market. In most Western industrialized countries, the steps from research to invention, development, or innovation to first possible use to refinement, implementation, or diffusion are usually distinct and disjointed activities; the steps are not continuous or interlocking. Furthermore, few, if any, decisions in the process are made by engineers. Therefore, new technology develops slowly and Western countries increasingly lose their competitive edge in the application or use of technologies, even if these are invented and developed in the same countries. Countries in the Far East, on the other hand, use much more continuous technology development and applications procedures.

Similarly, education in North America and Western Europe is largely concerned with pure science, the social sciences, narrow specialized professions, and general management, with the number of science graduates larger than the number of engineering graduates; both science and engineering graduates are far outnumbered by social science, humanities, and general management graduates at both the undergraduate and graduate level.

Japan, on the other hand, graduates more than twice the number of engineers per capita than the United States and four times as many as Western Europe. Similarly, in the United States engineering is generally less respected than other professions, such as medicine, law, and banking, while the reverse is true in Japan and most Asian societies.

One of my purposes here is to study the declining role of engineers in the leadership of the Western industrialized countries in North America and Europe and to suggest changes that could result in a reversal in the decline of technological and economic leadership of these countries. Society's perception of engineers and the function of engineers in advancing society's objectives are also discussed.

The idea is advanced here that societies must not become only users and consumers of technology but must learn to understand the role, basic workings, and impact of technology. Only then will society become a rational user of technology. To achieve this goal, engineers will have to learn to educate and advise society in the choice and application of technology much like medical doctors increasingly are involved in preventive medicine and not just treatment.

This book has its origin in an undergraduate seminar on the role of engineers given at the Massachusetts Institute of Technology (MIT). The seminar was attended by engineering students who, though interested in engineering as a profession, were unsure of their future role in society and by social science, political science, and management students who found themselves increasingly dependent on technology and involved in technological decisions without adequate background and knowledge.

The discussions generated at the seminar demonstrated a concern with the lack of involvement of engineers in society at large, in technological decision making in general, and in the management of technology by government and industry; the discussions also highlighted an inadequate knowledge on the part of the students of the bounds, capabilities, and impacts of technology.

This book addresses these concerns, discusses their origin, and suggests solutions designed to move us toward a more rational, efficient, and concerned technological society, able to cope not only with its own real time problems but also with the maintenance of the earth's environment for future generations.

The book evaluates engineers as professionals and engineering as a profession. It suggests radical changes in engineering education, in the role of engineers in society, and in the functions of engineers in technology development—from discovery through innovation. New models for the engineer of the future in the work environment and in society are presented, and the requirements for engineers to function under these new conditions are discussed. The impact of these changes is reviewed, and the needs for improved management and communication skills by engineers are established.

Finally, the requirements for technological excellence are developed and the correlation with a renewed engineering profession and engineering leadership is established. It is argued here that technological excellence is not feasible without effective engineering leadership and that both are interdependent. These findings are summarized in guidelines for the achievement of technological excellence and the attainment of engineering leadership for use in high-technology countries, as well as in those that seriously lag behind technological development and application.

Without technological excellence and engineering leadership economic growth becomes elusive and difficult to attain. It is the purpose of this book to provide the rationale and approach to the development of sustainable economic growth through technological excellence and engineering leadership.

2

Issues and Problems of Engineering Leadership

While it is generally recognized that technological change is critical to the success and growth of nations, firms, and individuals, the management of technological change remains among the least understood aspects of government and management. The contradiction is that technological change at this time is credited as a major contributor to the advance of industries as well as the economic growth of regions and nations.

The factors that contribute to the effectiveness of the management of technological change that have received scant attention, probably because an adequate data base of performance of technological change under different government or management conditions is not available.

The role of engineers in the invention or discovery and development of technology is well understood, and engineers are expected to implement, operate, or use such technology. But engineers seldom are asked to participate in the decision making that affects the selection, timing, rate of introduction, and use of technology.

The role of technology in society is all pervasive, and the structure of society has been greatly affected by technological change. Developments in communications, for example, have reduced the need for face-to-face contact for most purposes. Technology is now available that allows us to see the person we talk to by telephone and we can transmit documents instantly by facsimile machine. Yet industry, government, and the judiciary all maintain their traditional face-to-face decision making approaches, even though these requirements and some of the regulations or adjudications connected with them have been made obsolete by technological change. (For example, computer-aided negotiations and computer- or expert system–designed contracting can greatly reduce costly and time-consuming face-to-face negotiations and assure more effective design of legal instruments.) Engineers generally contribute little to economic and technological policy decisions by industry or government even when the decisions are based on technological factors. Engineers, at least in Western societies, are assumed to be specialists who cannot make decisions that involve consideration of broader aspects or implications.

Engineers are perceived as hardheaded, rational thinkers whose professionalism, if it can be so defined, is narrow and specific. By comparison, "true professionals" as perceived by the public, such as doctors, managers, accountants, and lawyers, are generalists who have professional skills but learn or know how to apply them in a wider context.

TECHNOLOGICAL SITUATION

The competitive position of the older industrialized nations, like those in Western Europe and more recently the United States, is deteriorating, notwithstanding increasing expenditures on research and development, education, and capital assets. A large percentage of inventions and discoveries made in these countries does not lead to new processes or products; in fact, the majority of inventions or discoveries are never even subjected to an innovation or development process. (Between

1950 and 1980, only 12% of patents filed by U.S. citizens, and corporations on their behalf, have been developed into useful technology.) Those that are often take years before being brought to application because of the slow progress in innovation and implementation, a lack of or insufficient support, the conflicting interests of the underwriters who often prefer to hide than to develop a new technology, and the perceived lack of interest by the market. A major reason for some of these delays is that in most Western industrialized countries, the steps from research to invention, development, or innovation to first possible use to refinement, implementation, or diffusion are usually distinct and disjointed activities; the steps are not continuous or interlocking. Therefore, new technology develops more slowly, and Western countries increasingly lose their competitive edge in the application or use of technologies, many of which were invented and often developed in these same countries, to the new industrialized nations in the Far East, which use much more streamlined continuous technology development and applications procedures. This situation, as discussed later in this book, is the result of differences in (1) the structure of society, (2) the perception of the role of technology, (3) the status of engineers and understanding of the function of engineers in an increasingly technological society, and (4) the acceptance of technology as a contributor to improvements in the standard of living. Most important, society is involved in technology not just as a user or consumer but as an advocate and supporter of all who contribute to technological and societal advance.

ENGINEERING TRADITIONS

Engineers like to work and make decisions under conditions of precision and predictability even though they are often aware of uncertainties in their measurements or assumptions. Physical problems that are the primary concern of engineers are assumed to require an accurate, systematic, and logical approach for solution, while managerial problems are often assumed to be subject to uncertainty, rapid change, imprecise definition, and a rather general method of solution. This traditional concept does not appear valid anymore. More and more management problems require a logical, analytical approach, while engineering is often done under great uncertainty and under conditions that require considerations of the larger social and physical environment.

Another trait often shared by engineers is the inability to communicate effectively, particularly with nonengineers. Engineers may be able to articulate their ideas or problem solutions in engineering jargon or technical shorthand to a narrow group of experts in their field, but in the Western world they often are unable to express themselves clearly in speech or writing. This difficulty may result in the ideas expressed being buried in obscurity. As a result, many good engineering ideas are not considered at all or only on the surface after a very long delay. The ideas are reinterpreted by others with better communication skills but who often have a limited comprehension of the issues involved.

Although engineers deal with changes in technology and therefore the future, few consider it their function to plan for the future or to consider the impact of future technological developments, even if they themselves are the instigators of such future technological developments. In fact, few engineers in the West understand the problems and methods of strategic planning. Fewer have ever attempted identification of technological voids and the forecasting of technology. In fact, engineers usually are the most cautious of forecasters and feel comfortable only in projecting engineering developments over the short run. Most important, they often have difficulty imagining the wider applications and implications of future technology.

ENGINEERING ROLES AND FUNCTIONS

The role of engineers should be changing rapidly now. Our increasingly technological society requires technical decision makers not only in the design and manufacture of engineered systems, processes, or products but also in the use and management of such systems. Technology is all pervasive and has infiltrated most activities, particularly in the industrialized countries of the Western world.

Different perceptions of the role of engineers are held, both by engineers and nontechnical people. These perceptions often are affected by cultural, historic, social, and environmental considerations. As an example, engineers play a role in Japanese, Korean, and other Far Eastern industries far different from that played in the United States. While over 80 percent of senior managers in Japanese industrial firms have an engineering background, less than 10 percent of managers in similar positions in the United States have an engineering background. The vast majority of senior marketing, planning, and production staff of Japanese industrial firms has an engineering background.

One of the reasons for this difference appears to be that in Japan the principal functions of management are strategic and technological with financial control considered an essential but consequential activity. U.S. management considers its principal function to be short-term financial management and control, an approach encouraged by American business schools. As a result, most managers in Japanese industrial firms advance from engineering, production, and operational positions to management. In the United States, engineers in these areas usually do not advance beyond their respective departmental structure, and company management is largely drawn from other disciplines (often from outside) such as accounting, finance, law, and marketing. As a result, management often is unfamiliar with the firm's technology (both process and product). Furthermore, the continuity of technological and financial decision making is impeded when top managers consider plans or decisions in financial terms and engineering and operating/manufacturing managers and staff are concerned principally with technological decisions.

Some claim that the differences between Japan and the United States in approach and in the role that engineers play are major factors influencing technological and productivity advances and account for much of the difference in industrial performance between Japan and the United States in recent years. While the differences may not be the principal determiner, they certainly affect the rate of technological advance, the effectiveness and timeliness of technological strategies, and, ultimately, economic competitiveness.

ACCEPTANCE OF CHANGE

While some claim that the U.S. government and U.S. industry are increasingly complacent and adverse to change and the inherent risk in change, others claim that the U.S. approach is the result of U.S. and Western society's basic adversity to risk and its traditional perception of superiority. This attitude has been fostered by a long-term feeling of the superiority of Western culture, institutions, and, in turn, science and technology. In fact, the United States appears to have become increasingly satisfied with mediocrity and imperfect social ideals and appears to have accepted less than perfect technology out of ignorance of the history and developments elsewhere and out of a lack of interest in things not American.

In the past, U.S. government, industry, and the public have been very slow to accept new ideas. At the same time, the purchase of new products from abroad has accelerated as more Americans become dissatisfied with U.S. product quality. This foreign buying trend exposes Americans consumers to new innovative technology. To reverse this trend, U.S. business strategies must be created that improve public acceptance of innovation. U.S. scientists and engineers work around the clock to create new inventions and then work at tortoise speed developing and diffusing new technology only to be overtaken by foreign competitors, often in the development of the same technology or product. In addition, product technology invention and innovation must include cost cutting, quality management, and related process innovation.

Government policies usually are designed to support the old, often defunct, technologies rather that the rapid development of new technology. This type of distortion in policy and investment contributes to the problem rather than the solution.

Engineers, who have long been subjected to a risk-averse environment, are themselves a major contributor and perpetuator of this condition. Engineers, forever unsure of their social and management position, usually are even more conservative than the users or consumers of their creativity or effort. Engineers are among the most risk averse of professionals. They always will introduce additional factors of safety even after overdesigning a product or project. It is not that they do not know their job or do not trust their computations, but that they suffer from an excruciating fear of public criticism or leadership displeasure; as a result, they will take few chances, even if common sense tells them that no overdesign and safety factor will ever eliminate all risks.

The factor of safety, introduced by engineers, is seldom derived by a scientific method. It more often than not is the result of some bias introduced by experience or hearsay. For example, safety factors may add as much as 30 percent to the weight and cost of a ship's hull without measurable improvements in ship safety, reliability, or maintainability.

ENGINEERING DECISION MAKERS

Engineers in the West are seldom decisive, particularly if a decision is to be based on insufficient or nonverifiable information. Engineers usually assume a rather contradictory approach. On the one hand, they will be pedantic in their approach to data and its use in engineering design and will go to great lengths to assure an accurate answer (often to many more degrees of detail than the data justifies); on the other hand, they will multiply their painfully and accurately obtained result by some "safety" factor which usually has no basis at all in the data or even in qualitative experience.

Engineers will be able to say exactly how much a structure can support under certain constrained and idealized conditions but will have so little confidence in their own results as to grab a safety factor of 2 or 3 or more out of the air and multiply most, if not all, calculated scantlings (thicknesses) by this number. In some cases, the safety factor is derived from so-called empirical experience, but in most cases, and certainly in cases where a new structure or system is designed, it is simply a quantitative reflection of the engineer's risk aversiveness. While a particular safety factor may be justified in some parts of an engineering design, it is seldom justifiable to apply it universally to all scantlings—yet this is what is usually done.

It appears that engineers would probably be wiser to spend more time and effort analyzing the uncertainties affecting their design and less on computing requirements for an idealized design operating under hypothetical conditions (see Appendix A).

INDUSTRIAL LEADERSHIP

Investments in fixed assets in Western industrialized countries are usually justified by short- to medium-term returns and are designed to meet market demands or counter competitive pressures. In other words, such investments are driven by short-term demand and profit and not by technology. In the Far East, by comparison, short-term technical obsolescence and projected long-term market changes are major forces influencing such investments.

In the United States, the world's largest industrial economy, nearly 50 percent of all scientists and engineers work for the federal government which, together with the defense and space industries, accounts for over 75 percent of all scientists and engineers. The rest of U.S. industry, private research organizations, and so on account for under 20 percent of the total, with about 6 percent employed by academia. Yet the bulk of all inventions in the United States are not made by the federal government, the defense or space industry, or even academia but by other, usually small, companies or research institutions. Even the large, mature, traditional industries—such as steel, textiles, automobiles, chemicals, and so on—account for a disproportionate number of scientists and engineers in relation to the number of inventions made or innovations introduced. It is by and large the small, often entrepreneurial companies (or individuals) that account for the vast majority of new inventions and ultimately innovations and technological developments in the Western world.

In the Far East, industrial firms, academia, government, and research laboratories cooperate much more closely. The problems the United States and other Western industrialized nations face today are twofold and involve the management of technological change and the management of "people change." Managers in the United States are trained to manage money but usually have less competence in and concern with the management of technology or people, tasks that are commonly left to the responsibility of lower-level managers. In the Far East, technological and people management are the principal responsibilities of top management, usually composed of professional engineers or technologists, while financial management and administration is delegated to a lower level. This approach permits greater flexibility in adapting training, incentives, promotion, and organizational structures to changes in technology and eliminates much of the disjointed, often uninformed and belated, personnel and technological decisions made in the West.

The management of technology today drives marketing, market share, and ultimately a firm's performance, if properly integrated with people management, and both provide major challenges and opportunities. Technology changes the working environment, work content, work condition, and, obviously, output, which in turn affects the value of worker output. Technological change nearly always results in fewer but much better jobs and the generation of new business and job opportunities which ultimately more than replace jobs lost by the initial technological change. To take advantage of the new job opportunities may require training of replaced workers.

The most important task for management today in government, in industry, and in service organizations is to learn how to manage technological change and the resulting challenges to strategic, economic, financial, material and human resource management. Unless decision makers acquire these skills and learn how better to involve engineers or technological experts, the future growth and competitiveness of Western societies may be at risk.

3

Societal Effects of Engineering and Technology

Technology is usually advanced to improve the quality of life and the standard of living, yet the most difficult problems facing humankind are not purely technological problems; they are social, economic, and political ones, although most are the consequence of technological change. While engineering or scientific solutions have helped resolve some social, economic, and political issues, many advances in technology have actually aggravated such problems or caused new ones. This often happens because technology is introduced without a full understanding of its implications.

The population explosion of the past 40 years, for example, is largely the result of improvements in nutrition and health care technology, yet at the same time living standards of the poorest nations actually have declined. While enthusiasm for technological and scientific discoveries and advances abound, such enthusiasm is now increasingly tempered by doubt, fear, or even the rejection of new technology or engineering solutions.

Such rejections often are based on a lack of information, wrong perceptions, bias, or simply inappropriate assumptions of how a technology functions or what its effects are. Most important, they often are the result of a lack of involvement on the part of the experts, particularly engineers, in the evaluation of technology and the subsequent decisions on the adoption and use of technology. This lack of involvement is largely the result of the reluctance of engineers to involve themselves in larger societal problems.

The environmental impact of technology is a typical example. Engineers usually take a rather parochial, or narrow, view of environmental impact and often consider only direct or primary impact with an emphasis on physical phenomena. Conversely, the public or society, unfamiliar with much of the technical and economic justifications for a new technology, usually will be concerned with secondary indirect impacts. Few engineers have learned or are able to consider technology in these dimensions.

Policy and decision makers in government or industry are largely untrained in or conversant with technology. Yet they bear the responsibility of formulating policy and regulations that affect technology development and use. These people usually consider technology from a macroscale and exogenous point of view, and they formulate their views and resulting decisions mostly on the basis of assumed effects or impacts, while ignoring how technology actually performs and relates to problems on a microscale. Yet technology today affects how we do things and what we do in many ways. It influences our choices and way of life. Unless policy and decision makers learn to understand technology, they may no longer be qualified to formulate policies, even macropolicies that invariably affect microissues directly or indirectly.

One reason for policy and decision makers' apparent complacency is that the societal and environmental effects of technology often occur or are noticed only long after a technology is introduced. Another reason is that policy and decision makers have been trained to use macroeconomic measures of economic and societal status or growth, even though it has long been recognized that such indicators of the state of an economy omit the effects of technological change. Today technology is the major force driving economic growth and world trade. The role of engineers in technology change or policy decisions is often undefined or confined to that of advisors or consultants; they are seldom involved directly in decision making. Engineers traditionally have hidden behind narrowly focused issues, and they do not employ their technological knowledge to help resolve major societal concerns. While it is true that society itself often avoids facing up to the need to resolve major issues such as energy use and environmental protection, engineers who often have at least some of the answers seldom take a lead in helping society resolve its problems. In fact, engineers often offer technological fixes that permit society to skirt the difficult decisions or issues.

As a result, some of the most important problems—such as the disposal of nuclear waste; the greenhouse effect, caused by emissions of carbon gases into the atmosphere; ozone depletion; ocean pollution; and more—either are not addressed at all or are skirted. Solutions to the problems often are deferred to a time when the problem is so magnified that rational solutions are impossible.

Engineers must deliver the promise of science and technology. In a way, they are the implementors who convert scientific discovery into useful products and applications. They should be aware of societal needs and societal limitations; with their knowledge of scientific principles and scientific constraints, they should develop technological solutions that satisfy society's demands (see Figure 1). This requires technology to be developed for use to imbed technological feasibility and societal needs.

Engineers must be aware of the effects of their work on society both directly and indirectly. They must have a perspective of how their work impacts on society, including the potential use or misuse of their creations. They should be able to make realistic projections of the advance of science and technology, where the latter is a reflection of human demand. This demand is often driven by unrealistic or even exaggerated expectations. Engineers are responsible for setting realistic technological projections and for defining the feasible scope of technology. They may have to tone down perceived societal needs, or they may be able to advance them in light of feasible technology that may offer greater technological opportunities than society expects.

Figure 1
Societal Needs, Demands, and Technology Application and Use

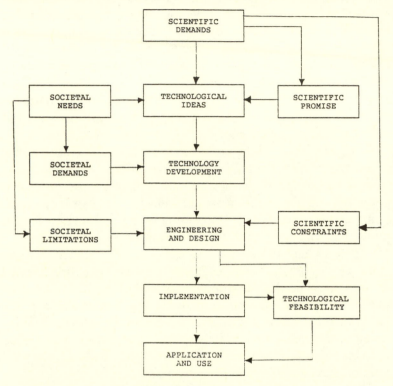

Engineers, therefore, are the ultimate applicators of scientific knowledge in the development of technology and the solution of problems. The process is a push-pull one in which engineers increasingly play a proactive role. As such, and as the purveyors of knowledge on technology, engineers should assume the role of risk, opportunity, and performance evaluator. Similarly, they should evaluate the costs, risks, and benefits of new technology and provide leadership in technology policy decision making.

Society usually does not face up to realistic problems, as people in general are unwilling to confront the big picture. They are usually content or more comfortable with a narrow problem focus.

Engineers are no exception even though they usually understand the larger implications of a technological solution to a problem. They also prefer to focus narrowly and thereby try to escape from considerations of larger, more complex issues. Engineers also consider hazards and regulations as the negative side of technology which, if possible, should be brushed aside. To confront techno- logical decisions, engineers today must be willing to get involved in politically charged issues and ultimately become politically active.

Engineers must be prepared to assume greater leadership in the solution of societal problems increasingly caused by technology. If they do not, or if they shun this responsibility, they will remain skillful, narrow professionals who have renounced responsibility for the products of their endeavors.

THE MEANING OF ENGINEERING

The word *engineer*, which has its origin in the French or German word *ingenieur*, a word derived from *ingineous*, has in the English language assumed a somewhat different connotation than the Old World original. Now, *to engineer* means, at least in colloquial English, to manipulate or develop some solution. It is not neces- sarily a complimentary expression. For example, we engineer escapes of prisoners, illegal transfers of money, or any number of shady activities that require somewhat complex manipulations. In fact, *to engineer*, in Western and certainly English usage, often has a derogatory connotation. Lawyers will engineer changes in the status of illegal immigrants or in the bankruptcy of individuals or companies. *Engineering* and *to engineer* have become general expressions in the English language denoting the fixing of some- thing, not necessarily employing the most effective or honest means. It is difficult to determine the origins of the use of the word *engineer* as a verb describing somewhat questionable activities, and it may be difficult to reverse this usage in the short run.

Yet it appears that this change in the colloquial usage of the term was the result of changes in the status of engineers and the respect in which they are held. We now have garbage collectors who are called sanitation engineers, and boiler operators who are called power or maintenance engineers.

It appears that the term *engineer* has become somewhat fungible. It denotes or could be applied to anyone who performs any function in which some technology is used, maintained, or designed. In other words, society has a very vague idea of what engineers do or are trained to do. As a result, *engineer* has become a rather vague concept, unlike the title used for other professionals. Anyone could call himself or herself an engineer of some sort, at least in the United States. Professional titles in law, medicine, and accounting, however, are very closely guarded and are regarded as titles that legally entitle a person to work or provide services in a particular professional area. While these professions have been able to define the specific functions, roles, and responsibilities of their professions and have established well-defined boundaries that they maintain as their professional prerogatives or reserves and that are guarded by legal constraints, engineers work in a widely undefined environment that never has been regulated effectively in terms of professional licensing or otherwise.

Not many graduate engineers in the United States go through the motions of qualifying as a "professional engineer," in part because "professional engineer" examinations or qualification requirements are based largely on archaic craft and not on professional engineering requirements. Many professional engineers' state examination and licensing boards include members who have little or no formal engineering education. Similarly, many engineers find that the license or lack thereof does not impede or further their professional activities. This situation may be due to the fact that most engineers work for large firms and not as individual professionals. It is interesting that neither legislators acting on behalf of their constituents nor engineers as a group have exerted any pressure to change this situation, even though professional status and social as well as pecuniary rewards are closely linked.

In some Western countries, such as Germany and France, things are different and engineers require and qualify for legal professional titles that allow them to perform specific engineering functions. The same applies to some degree in certain Latin American countries.

4

Engineers and Leadership

Leadership is not inherited and it is not just part of personality; it can be acquired. People can learn to be leaders, but leadership qualities are not acquired equally. Leaders are distinguished from others by the way they behave, by their appearance, and by the way they act. Leaders must have vision to look into the future, forecast or project results, and imagine or conceive developments. They have breadth and depth of knowledge and the humility of admitting the limits of their knowledge. Leaders continually strive to expand their knowledge and competence. They believe in others and see their work as a mission.

They can envision outcomes and define objectives and goals, as well as define the knowledge required for the achievement of such goals. Leaders are good organizers who can convince others of the importance of the goal and participation toward that goal. They have a sense of personal commitment and duty backed by personal, unselfish responsibility toward achievement of the identified goal. They are willing to risk professional and social reputation by sticking to what they think is right, and they are willing to assume the penalties and consequences.

Leaders are able to apply their intelligence, experience, and training toward the achievement of a goal without losing sight of the economic and social consequences. Leaders are effective yet realistic planners, capable of communicating their visions and plans for the achievement of a goal. They are good managers who influence the supporting actions of others by example and who inspire as well as motivate others toward the achievement of a goal. Leaders must be persuasive in words and action and forceful in their example.

Leaders show undivided loyalty to their goal, organization, and coworkers and present integrity to self and others by practice of the highest ethical standards. Leaders persevere yet always show resourcefulness to make do with available resources. Leaders resolve problems and do not hesitate to redirect actions if it may lead to a more effective move toward a goal.

They show judgment and continuously learn from their experience. They are proud of accomplishments but give proper credit for others' contributions toward a goal.

For engineers to lead, they must learn to take broader perspectives in which engineering serves as one of many aspects leading toward a particular goal or solution to a problem.

Engineers often lack the personality for or approach to effective leadership. Leadership requires interpersonal and communication skills that engineers often lack; it requires an understanding of people and a knowledge of how to get along with them which engineers often consider unimportant. Leaders are sensitive to the feelings of others and know what to say under various conditions and how to relate to different people.

Leadership skill includes effective speaking, attentive listening, sensitive interaction, diplomatic interpersonal relations, tact, and compassion under appropriate circumstances. Leaders also are careful in criticizing anyone, particularly when they know that the other has tried his or her best. Encouragement and positive valuation of the opportunities are much more effective in moving others toward better performance without self-recrimination and loss of face.

People are motivated by positive approaches; if a leader cannot make a positive comment, under regrettable circumstances, then the best approach is to be silent. The perpetrator knows perfectly well that his or her performance was substandard and will be encouraged to do better by such a positive or neutral approach.

If a fault is to be discussed, it should be discussed in general terms and as a learning experience, not as an exercise designed to assign blame. The objective should be to prevent a recurrence of the problem and not to punish the guilty. Unfortunately, many engineers feel a responsibility, and often a need, to criticize or correct others, particularly other engineers, even if things cannot be changed.

THE CHARACTER OF ENGINEERING LEADERSHIP

Engineering leaders must show creativity, imagination, and an ability to present a vision of how to implement their ideas. Visions are new ideas, methods, and approaches that have not been tried or used and that make sense, are reasonable, and present a solution to a real problem. A leader's vision must be possible and acceptable to reasonable people who recognize it as a creative solution that is easily understood. Leaders must be able to communicate their ideas in simple, convincing, and logical terms. They must convey the integrity of their vision by showing its benefits, while at the same time convincing others that the vision is a worthwhile one.

Leaders and their visions must instill and project confidence and continuously earn the trust of their peers and staff. To be effective, organizations need many leaders and not just leaders (or a leader) at the top. Leadership is not developed by appointment or promotion; it is developed with creativity, a sense of responsibility, an understanding of others, and the ability to communicate with others at all levels.

Leaders are committed to ideas or solutions not because of their origin but because of their value and relevance to the objective of the entity (or public) they serve. Engineering leadership characteristics include the following:

1. Doing one's best

2. Thrive on challenge

3. Attention to detail and perfection without losing track of total requirements and overall objectives

4. Knowledge and expertise

5. Courage and truthfulness

6. Persuasiveness and inspiration

7. Forcefulness

8. Integrity

9. Thrill of and satisfaction with achievement

10. Creative pleasure

11. Determination

12. Winning attitude

13. Intelligent risk taker and someone who encourages risk taking

14. Intelligence

15. Economic sense

16. Responsibility

17. Self-esteem and self-confidence

18. Talent

19. Discipline

20. Empathy and support

21. Loyalty

22. Competitiveness

23. Commitment

24. Resourcefulness

25. Perseverance and patience

26. Pride in own and others' achievements

27. Judgment and fairness

28. Communicator

29. Admits and knows limitations

30. Trust in subordinates

31. Visionary and able to instill faith

32. Admission of lack of knowledge

33. Able to simplify

34. Will to succeed

35. Motivated

36. Does not fear dissent and learns from mistakes and failures

Most engineers are not used to leading nor are they trained in or accustomed to the concept of leadership. When put into a position of leadership, they will often ask the following:

- How does one lead?
- What does leading really mean, particularly when leading people is involved?
- How does one express leadership in terms of ideas, motivation, example, competence, and rationality?
- What is the role of leadership?
- What is the responsibility of leadership?

Other questions often concern the technical component of leadership in terms of its engineering content. In other words, engineers may ask if it requires leading with advanced engineering inputs.

To be leaders, engineers must induce loyalty, goal orientation, good communication, interpersonal cooperation, and effectiveness. They should encourage motivation, perseverance, quality, and collegiality and they should assure effective progress by the use of planning and good organization. Most important, they must provide an example of integrity and ethical behavior.

Leadership requires more than training; it requires a leadership attitude. Engineering leadership skills can be acquired only by experience in leading, not by artificial or simulated problem solving. Leadership requires teamwork and practice in guiding group decision making. It includes an ability for self-evaluation and self-criticism. While leadership skills can be taught in engineering education, such a curriculum also must train engineers in understanding the context of engineering.

THE PUBLIC MAGE OF ENGINEERS

A report commissioned by the National Academy of Engineering (NAE) found that the "American public holds engineers in high esteem, but has only a vague idea of what they do." The report, prepared by the Public Agenda Foundation, was based on a series of "six focus groups," five with college-educated adults and the sixth with congressional aides and staff. Focus groups involved a small group of subjects, carefully chosen to be representative of various viewpoints, who had in-depth discussions of an issue. The report was the second part of an NAE project on public awareness of the contributions of engineering and technology.

In the first report, the Public Agenda Foundation summarized the findings of existing surveys on public attitudes toward technology and engineers. Many of those surveys revealed attitudes substantially similar to those held by the college-educated focus group participants.

According to the two reports, public attitudes toward engineers, engineering, and technology include the following:

- Engineers have an unusually high degree of integrity and are generally not to blame for technological accidents, mishaps, or disasters such as the explosion of the space shuttle.
- Policymakers and government officials should hear more directly from engineers about the development and use of new or risky technology, but the technical experts should not have the final say in whether to proceed with the technology.
- Engineers typically cannot communicate well with non-engineers, and many have "poor social skills."
- Engineering is a desirable and high-prestige career, and most people would not mind if their son or daughter married an engineer or became one themselves.
- American engineers, with the possible exception of auto engineers, are the best in the world, possibly exceeded only by the Japanese.
- Americans have great faith in technology and believe it has done more good than harm.
- Legislative leaders mistakenly believe that Americans define "new technology" in terms of nuclear power or the arms race. The public definition is much more mundane: VCRs, computers, and other innovations that change daily life, often for the better.

SELF-ESTEEM AND ENGINEERS

It is only recently that self-esteem has become an issue in education and the workplace. In fact, many experts now consider self-esteem as the dominant educational theory underlying design and performance of educational programs. Though some think that the rising emphasis on feelings comes at the expense of subject matter and may degrade the "quality" and "scope" of education, there are many who feel that effective education and professional work requires personal involvement, satisfaction, and, ultimately, self-esteem and self-liking. Some self-esteem programs use simplistic approaches based on simple, satisfying concepts and on the idea that people should not be too hard on themselves and should like themselves. Obsession with self-esteem often undermines a professional's concern with education and improvement. Real self-esteem depends largely on recognized, or at least a sense of, mastery. Engineers, as professionals, usually are proud of their work and their understanding of physical processes and physical requirements of systems, structures, and devices. They are proud of achieving designs that work and effectively performing their function. Yet engineers look at their accomplishments largely from an engineering point of view; they have little patience with others who do not comprehend the technical beauty, originality, or effectiveness of an engineering solution and who judge it from only a viewer's or user's point of view. The perceived lack of credit and valuation by others affects engineers' self-esteem.

ETHICS AND PUBLIC RESPONSIBILITY

Ethics, which is often associated with morality, is difficult to discuss, particularly when applied to engineers, who perform certain functions that are not always well understood. While we recognize the contribution of technology to the quality of human life, we increasingly are aware of damage—particularly environmental damage—caused directly or indirectly by technology. Consequently, engineers, as the developers of technology, often stand in direct conflict with public opinion and with some standards of moral behavior.

In many societies and throughout history, attempts have been made to legislate moral and ethical actions, yet most of these have been criticized or condemned as interferences with human freedom. While ethical standards usually are set to improve societal and interpersonal behavior, they often lead to rigid codes of behavior that do not benefit society. They may result in causing inconvenience or outright damage to some or all in a community.

Ethical and moral standards often are imposed by law or regulations that define both standards and responsibility. Unfortunately, standards sometimes are set by people unfamiliar with the state or capability of a technology, and responsibility often is assigned to people without perfect knowledge of the technology involved. For example, although it would seem obvious that the engineer who designs a product should be responsible for the product's safety, the engineer usually is not responsible. The problem of safety involves several issues, as follows:

1. Safety manufactured into the product or process

2. Material choice in the manufacture of the product

3. Use for which the product is marketed

4. Directions for uses developed

5. Controls on product distribution and use

6. Product grouping or segmentation

Engineers seldom are involved in or consulted on any of the above issues, particularly in the U.S. Their job is assumed to be finished with the completion of the product design.

Although the quality and safety of products usually can be improved at little, if any, added cost, engineers in the United States and Western Europe often are not involved in decisions that affect these concerns. In most cases, admonitions of engineers are ignored, and engineers become involved only after an incident or, worse, an accident involving a product.

Over many years engineering societies and similar professional associations have published codes of ethics (see Appendix B), both for the individual engineer and for engineers who are members of corporate or government peer groups. A problem with these codes, all of which are designed to oblige engineers to maintain high professional standards as well as standards of conduct among their members and in their relationships with the public, is that they usually are defined in narrow professional and legal terms. They usually ignore moral, including environmental, issues.

Another problem with the codes is the near total lack of enforcement. Some societies advocate peer review as a method of enforcement, but most often it is left to legal enforcement to correct radical departures from safety or other standards. The lack of formal procedures and only after-the-fact enforcement are major problems engineers must face. The space shuttle *Challenger* disaster is a typical example of this problem. It highlights the problems of whistle blowing by responsible engineers.

Whistle blowing has received much attention in recent times, though the courageous engineers who have come forward to identify unethical behavior, bad engineering, lack of safety consideration, and even fraud seldom have been recognized or rewarded. In fact, their lot more often is loss of job, loss of reputation, and loss of social standing.

ETHICAL STANDARDS FOR ENGINEERS

Although most professionals, including engineers, are supposed to be guided by firm ethical standards, they often are not. The public has become concerned about the actual application and use of such standards. Engineers have been isolated from some of the recent uproar because engineering failures are

1. more difficult to understand and interpret,

2. more difficult to identify,

3. more difficult to pinpoint, and

4. more difficult to associate with a particular person. It is only when a major technological disaster strikes, such as the failure of a pedestrian bridge in a hotel or the *Challenger* disaster, that the public asks for answers.

In the past, engineers had not been identified as targets in ethical inquiries because they seldom assumed ultimate decision-making responsibilities. Yet, in this age, where products and processes embody ever-increasing technological complex-ities and potential hazards, engineers must assume a greater role in responsibility for public safety, environmental acceptability, and social compatibility.

Engineers generally have not been accepted as responsible decision makers who provide the technical expertise pertaining to the impact of technology on the public. In the *Challenger* disaster, for example, managers disregarded the advice and warning of engineers regarding safety during the launch.

By and large, engineers are not trained as decision makers or responsibility takers. Furthermore, they often are given responsibility for only a part of the whole picture; as a result, they do not have an opportunity to comprehend fully or judge the implications of their decisions. Nonetheless, engineers have a responsibility not only to themselves and their employer but also to their peers, the users of their engineering solutions, and the public at large.

ENGINEERING ETHICS AND MORALITY

Engineering ethics can be considered objectively, subjectively, or by the use of personal standards and concepts. In an objective approach, we assume that right and good are objective properties of behavior that guide engineers. In the subjective approach, we assume that engineers refer to their individual feelings of right and good. In the comparative approach, we assume that engineers consult some standard or concept of behavior. Though few engineers consider ethics explicitly, their behavior, as well as their approaches and responses, is guided by ethics. Similarly, the rules of ethics engineers employ are seldom thought through. More often they are used without any plan or awareness.

Ethics and its rules are influenced by moral principles which are affected by culture, religion, upbringing, and, increasingly, the environment. Much of ethics has to do with the standards of acceptance developed by an environment and reinterpreted by the individuals in a community.

Engineers often face the dilemma of a conflict among fundamental or moral rules of ethics, rules set by laws and codes, and standards of behavior provided by peers and professional groups and associations. As a result, engineers sometimes are subjected to conflicting pressures of

1. doing the right or moral thing,

2. providing the best engineering solution,

3. assuming absolute safety and reliability,

4. solving the problem in the most efficient (often interpreted as economic) manner,

5. considering environmental and cultural traditions,

6. following set codes and standards that may or may not provide an effective approach,

7. obeying professional procedures that may not necessarily lead to an effective solution, and

8. considering political influence in engineering decisions.

Such pressures routinely are exerted on engineers who must somehow balance their professional approach with satisfaction or rejection of such pressures.

5

Engineers as Decision Makers

Few engineers reach positions of leadership in the West, and when they do it is seldom the result of their superior performance or standing as engineers. Engineers generally are assumed to have few management skills and to be indecisive. As professionals they are seen as unable to manage their own public or intra- and interprofessional relationships. They also are assumed to have poor communication and marketing skills and to lack assertiveness.

Management and engineering are interdependent and not mutually exclusive activities, even though management often is general and engineering specific in nature. The public perception is that managers are generalists able to manage anything, while engineers are indecisive and suffer under management paralysis through overanalysis. Engineers are assumed to be unable to consider broader aspects affecting management decisions.

Business schools have contributed greatly to this perception. They have, for years, emphasized the rationalist and bottom-line approach to management, which, to some extent, has led U.S. industry down a dangerous path of short-term optimization and long-term stagnation. Management today often lacks an effective perspective and identification with the company and its products. It is often isolated and not people- and society-oriented. For engineers to help management overcome its problems, they must become better decision makers, more assertive, improved communicators, and team players.

ENGINEERING GOALS AND VALUES

In their work engineers usually employ narrow technical objectives, often defined by weight or cost of structural materials used. Their focus is largely the result of their narrow technical training. Although most engineering colleges emphasize humanities requirements, non-engineering offerings usually are taught as distinct subjects which are supposed to induce some culture but which are not designed to affect the engineering approach or to equip engineers to consider broader issues of society and culture in developing engineering solutions. In fact, few engineers recognize any relationship between cultural value and content and engineering. As a result, engineering objectives nearly always are pure engineering goals that can be achieved by the rigid application of engineering principles.

Engineering scientists often adopt the practice of asking if a new theory expounded to explain a technical phenomenon is logically and scientifically correct, but they often evade answering the question if the new theory helps to understand observable technical or technological phenomena and, ultimately, to solve a real-world problem. In other words, the goals of many engineers have become narrowly technical. To be accepted as decision makers, engineers will have to show that they know how to consider the larger implications and that they are concerned with the solution of real problems and not only with the development of technically interesting solutions.

REGAINING ENGINEERING LEADERSHIP

New directions must guide Western governments, industry, and society if they are to regain engineering and thereby economic leadership. We must train engineers to assume not only engineering but development decision making responsibilities. Engineers must become involved in setting technological policy at the government, industrial, and societal level. They must be educated to act as leaders and managers if we are to maintain our economic position and standards relative to other advanced nations.

We must reevaluate our concept of the role of engineers in a technological society and reestablish engineering leadership by assuring that technological decisions are made by engineers with the assistance of legal, financial, and marketing professionals—and not the other way around. The United States never would have gained its economic and technological preeminence, lost only in recent years, if our engineering geniuses like Henry Ford, and Thomas Edison had been relegated to the positions engineers hold in U.S. government and industry today. To move in the necessary direction, we must do the following:

1. Use performance standards instead of technical specifications in government or industry procurement to allow the most effective technology to be introduced.

2. Use education budgets to redirect the U.S. educational system to assure future availability of the required professionals for the technological age for which we must prepare.

3. Assure that research and development (R&D) becomes an integral part of the innovation and technology development process and not a distinct once-through activity that does not communicate with the subsequent stages of technological development.

For example, to this day, the U.S. Navy insists on specifying each minute detail of every naval ship procurement. As a result, it takes from 4 to 8 years between the establishment of performance requirements and the start of acquisition of a new naval system, which then takes another 4 to 8 years to build or deliver. The added acquisition costs are phenomenal, but even more serious are the costs of built-in obsolescence.

U.S. primary and secondary education largely concentrates on preparing students for college. High school diplomas are assumed to be required for most career paths, and little emphasis is given to career paths that do not require a college degree. The percentage of students going to college in the United States is twice that of any other developed nation. Our average primary, secondary, and college education is inferior. Tremendous waste is generated funding this education, both in resources and in time. People who do not benefit from a college education waste 4 years when they could have used that time advancing their skills and contributing effectively to their life's objectives and the nation's economy.

The cost and waste amounts to hundreds of billions of dollars that could be spent more wisely in advancing the standards of U.S. primary and secondary education, introducing effective vocational training, and improving the general standard and professional orientation of colleges. In particular, resources will be required to advance the standard of mathematics and science education at the primary and secondary school level so that universities truly can become centers of scholarly activities and professional education and not largely remedial educational centers that teach students the basics of science, English, and other subjects that should have been covered in high school.

Research and development must be liberated from its ivory tower in government, industry, and academia, if it is to make a meaningful and timely contribution. We also must learn to value inventions, discoveries, and subsequent innovations with a long-term and not just a short-term perspective, as well as allow changes in innovation even when short-term payoffs are not immediately evident. For too long now, we have arrested many promising R&D efforts or discoveries because there was no immediate profit potential, only to see our major competitors pick up the ideas and develop them to their benefit.

Our short-term perspective not only has given other countries a competitive edge, often leading to market domination in our own home market and abroad, but also has led to a rapid decline of patents filed in the U.S. Patent Office by U.S. citizens. Although U.S. patents filed have increased very slowly by just a few percent over the last 10 years, foreign patents filed in the United States, which constituted barely 8 percent in 1976, reached a level of between 46 and 48 percent in 1990 and are expected to exceed the number filed by U.S. citizens by 1995.

Many foreign patents are based on U.S. scientific discoveries that were not developed further. Others are new applications of U.S. inventions. To regain our technological leadership, a prerequisite for economic leadership, we must give U.S. engineers a greater role and responsibility for leadership.

PRESTIGE AND STATUS OF ENGINEERS

In a recent poll asking high school students what engineers do, students identified auto repair mechanics, building superintendents, and locomotive drivers as engineers. Most of the respondents felt that these were not only typical jobs or functions of engineers but were the principal roles of engineers. Part of this perception may be the result of people taking much of technology for granted and not being aware that technology is the result of engineering. Yet engineering associations do little to reverse or even influence this public perception. As a result, few engineers achieve positions of leadership in government or industry.

ENGINEERS' INTERPERSONAL AND COMMUNICATION SKILLS

You can get an engineer to generate a fifty-page report overnight, but you cannot get him to develop a one-page analysis or a simple graph that outlines his or her recommendations and the reasons for them.

Most engineers lack effective interpersonal and communication skills. There are many reasons for this. The most important probably are that engineers seldom work directly for customers and that their work usually is highly focused. Furthermore, their education, as a rule, includes little in terms of communication, interpersonal skills, and behavioral subjects. Engineers are trained to work in small, highly specialized, and often isolated groups.

Engineers use symbolic shorthand in practically all of their communications. It is an efficient method to communicate with their peers but is incomprehensible to others. Yet few engineers have the patience or ability to translate their ideas into plain English. When required to do so, they often find that they do not have, or have lost, the skills for effective verbal communication.

Their interpersonal relationships often suffer, as they expect others to focus intensely on and to be interested in their particular expertise. They similarly may lack the patience for the social niceties that are essential for smooth and effective interpersonal communications.

Engineers often put little emphasis on external factors such as dress and social customs and consider these a waste of time and resources. It is interesting to note that these are issues greatly emphasized by other professionals, such as lawyers, doctors, and even architects.

MARKETING OF ENGINEERING

Engineering is the only profession that is not marketed effectively. It is not marketed to the general public, not to industry, and certainly not to decision and policymakers. We may see advertisements by associations of teachers, lawyers, and doctors, informing the public of the great contribution of these professions. We find that these organizations devote significant efforts to lobbying in capitals such as Washington, D.C.

These organizations consider their primary role to be marketing their profession. Yet little, if any, marketing is conducted by engineering organizations. While part of the reason for the lack of marketing may be the image fragmentation of the engineering profession, the base cause appears to be engineers' introversion and lack of self-confidence.

Little is heard of engineers in the media as well. There are no engineers in soap operas, and they seldom are consulted in the news or public interest programs. Engineers and engineering seem to be taken for granted, and few among the public see any excitement or drama in engineering.

It seems that engineers are not publicity conscious and that they are in some ways secretive. They are respected for what they produce, for the technology they supposedly develop, but not as individuals or even engineering organizations. How many among the public have ever heard of any of the engineering societies? Engineering is neither effectively marketed nor explained, and until it is the role of engineers will not be understood and their status distorted.

ENGINEERING CREATIVITY

Creativity in engineering is an enigmatic characteristic. Engineers are among the most creative and imaginative as well as the most traditional and hidebound of people. They will come up with brilliant ideas or solutions but will hesitate to introduce them because the approach has not been used before or because they fear that it may make them controversial.

Somehow engineering creativity and risk aversiveness collide, and tradition is the invariable refuge. Much of this is probably the result of the dichotomous standing of engineers as professionals without a clear definition of their role, function, or status. As a result, engineers often try to conform, to be more popular, even if they know that their untraditional idea offers a better solution.

While there are many myths of creativity, modern investigators seem to agree that creativity is not the result of sudden inspiration and flash of genius, but more often is the result of a highly logical process. Creativity requires focusing intensely on a problem, opportunity, process, or phenomenon; analyzing it; and then developing many alternative solutions or approaches. The analysis usually is based on known theories, facts, or experiences. The creator consciously or unconsciously describes various alternatives systematically, on the basis of prior knowledge, testing assumptions and constraints repeatedly.

A good creator must be steeped in the area of the subject and have formal and informal experience in it. Flashes of creative ideas usually are generated by focusing on a difficult or interesting problem. Creation seldom occurs and most certainly rarely matures into real breakthroughs or opportunities without hard work, commitment, and an ability to see the creation in perspective.

Engineers are expected to be creative because of their training. While most people seem to be able to recognize creativity, few can define it in an effective rational manner. Engineers are supposed to have the background and knowledge to be creative, yet most are too narrow in their perception of problems, and their solving ability, to be able to create effective innovative solutions to problems.

Engineers usually show technical creativity, which means they are able to show new or different means of solving a narrow, well-defined problem that is not subject to environmental or user uncertainty. They are good, for example, at creating a new method of fastening a defined load under given structural and environmental conditions, but would be less creative if given a vaguely defined problem with lots of ambiguity and uncertainties. As a result, most, but fortunately not all, engineers are marginal creators. They are good at creating marginal improvements but often fail when radical innovations are needed. Scientific engineers often create marginal innovation in scientific or technological theory; others will develop new devices, systems, structures, or processes.

In the United States, the innovation process is highly concentrated during the research and invention stage, and large numbers of engineering scientists are used to further the process. Engineering involvement, and often interest, falls off during the innovation and final application stage. This circumstance is quite different from that found in Japan, where, for example, engineering creativity is concentrated increasingly toward improving the technology innovation and application process (see Figure 2). In other words, the Japanese are not concerned mainly with invention but with

Figure 2
Employment or Emphasis of Engineering in Technology Development

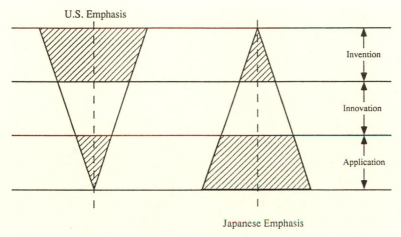

U.S. Emphasis

Invention

Innovation

Application

Japanese Emphasis

effective creative engineering of the innovation and improvement as well as ultimate application and use of a new technology.

Engineering creativity does not end with a technological breakthrough; it is a continuously accelerating process that stops only when a technology becomes mature or obsolete or no longer has a market. Figure 3 indicates the rapid growth of investment in engineering in Japan during much of the innovation stage. It also shows the continuity of research and innovation by the merging of innovation engineering with research and development by scientists and engineers. By comparison, the U.S. experience is more disjointed. Research and development is performed by distinct groups of scientists and engineers and is followed, often after a delay, by a slow and often hesitant innovation effort which often peaks long before a technology has reached its potential.

Creativity is characterized by originality and imagination. Being creative involves developing original ideas. One area in which technical creativity should play a role is in the development of new devices, systems, and structures, the area in which the vast majority of engineers (designers) work. Here the task is to apply scientific,

Figure 3
Engineering Role in the Innovation Process

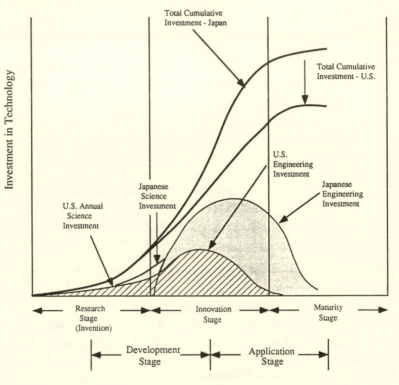

mathematical, economic, and social knowledge to satisfy specific needs. In order to create new devices, systems, or structures, the following steps should be taken by engineers:

1. Recognition and definition of problem at hand

2. Development of some outline concepts

3. Consideration of alternative approaches to solving the problem

4. Decision on a method of solution

5. Designing the solution

Technical creativity, therefore, entails not only imagination but also good analytical skills, an ability to consider a problem's wider perspectives, and the whole range of alternative solutions.

While some people appear to be natural creators—that is, ideas seem to come forth almost automatically—others have to engage in a deliberate, conscious effort to develop their creative potential. Studies indicate that most persons (engineers) have far more creative capacity than they generally use. That people use less of their capacity is due largely to structural, organizational, and motivational impediments.

Brainstorming may be an effective method for enhancing creativity, particularly for engineers who need a peer group as a sounding board in order to verify that their ideas make sense and are implementable. Similarly, ideas can be refined effectively in brainstorming sessions. This group interaction has become an effective approach toward problem solving and creative idea development.

Concentration is another method of stimulating one's creative capacities. One way or another, one must get to thinking, long and hard, about possible solutions to the problem and developing as many leads as possible. Creativity is a long-term process that cannot be effectively planned and budgeted. Fostering creativity, therefore, requires a long-term view, which few U.S. companies adopt. U.S. companies must lean toward long-term philosophies in order to compete aggressively with the Germans and Japanese— no matter what the industry.

6

Managing and Organizing Engineers

Engineers are a breed apart. They are unique and capable but are forever trying to find their position in an organization or in society. As a result, engineers often have difficulty fitting into line organizations.

Managing engineers require systematic nurturing of their technical interests to be effective. More focused human resource management is required than is usually expected. Engineers require not only consideration as individuals but also an environment that fosters creativity, networking, and collegiality. Most important, they require an environment that provides incentives for creativity and that encourages the development of human potentials. The environment must stimulate and, therefore, should not be very structured. Most engineers are greater specialists in some areas than in others, and it usually is difficult to determine which specialty or knowledge is most important. The relative importance of engineering problems often will change from job to job; an engineer whose contribution to one job was not very important may find that he or she serves a crucial role in the next job. Engineers must feel that they are respected and that their work or contribution is recognized. Engineers require more encouragement than other professionals because they work more often on virgin problems and must be given confidence that it is all right to work on some new approach and to make mistakes. Mistakes are not a reflection on competence but are a positive sign that the engineer is willing to take chances, be creative, and assume risks as a necessary step in technological progress.

Engineers become frustrated if they feel they are not being utilized effectively, and the principal reason for engineers' job dissatisfaction and departure is often a feeling that industries will not use all that engineers can contribute. U.S. companies often do not have a corporate culture that is conducive to the effective utilization of engineers. Engineers' frustration is, to a large extent, a result of the following:

1. The role engineers play in the corporate structure

2. The perception of the contribution of engineers who are often considered technical hired hands who can readily be replaced

3. The lack of opportunity for engineers to advance or to fill important decision-making positions. To advance, they must usually depart from their engineering roles

4. Discouragement of networking and the lack of effective internal communication channels

5. Lack of opportunity for engineers to learn about manufacturing, planning, marketing, and so on

6. Lack of feedback from marketing and strategic management to engineers

7. Lack of opportunity for self-development. While managers or sales staff are encouraged to attend seminars, courses, or meetings, continued training of engineers, particularly midcareer retraining, is usually not considered essential. Management ignores the experience of midcareer engineers and

hires young engineers with newer skills; the experience of older engineers gets lost in the process

8. Ineffective reward or recognition systems, with ill-defined career paths, which often require engineers to change their affiliation to management or marketing if they wish to advance. This is true in both industry and government

These and other factors contribute to the feeling of engineers that they are largely outsiders, not part of the establishment peer group. Industrial organizations and their management of engineers will have to change in a very drastic way if they are to hire and keep the people needed to create tomorrow's product, process, or service technology.

Many engineers are disillusioned with their jobs. Loyalty and performance decreases, as they feel less excitement about their jobs, less meaningfulness in job content, and decreasing opportunities for vision and creativity. Many feel that the role of engineers should be reevaluated as engineering work functions change. Engineers are less required to perform design and more to integrate scientific and engineering solutions. There is a driving need for problem solution and less of a need for microdesign which is effectively done by computer-aided or performance design tools. Yet engineers usually are not given the responsibility or authority to make larger or integrative decisions and to become systems integrators. They feel that management actually mistrusts them or at least feels that engineers cannot be trusted with "larger" decisions. As a result, and at a time when companies need greater involvement of engineers, many engineers feel less involved with their companies. (See Kanter and Mirvis, 1990).

A related issue is the compensation that engineers receive. They usually earn a small fraction of the salary that managers earn, who further benefit from costly fringe benefits. The average total salary of senior or executive engineers in U.S. manufacturing firms was less than one-third to one-half of that of upper-middle-level line managers in 1990. In other words, a senior or managing design engineer responsible for a new product which could make or break the company usually will make only one-third the salary of the manager of marketing (or sales), the treasury (or accounting), or the legal department. The gap becomes even larger when the cost of fringe benefits is added. For engineers in the United States to overcome this gap will require a change in education, attitude, and, most important, self-perception.

As stated, to be effective engineers must be nurtured. Motivating engineers is more complicated than motivating other workers or professionals who may be motivated by direct monetary, status, and other recognitive awards. Engineers usually demand also that the job be interesting, meaningful, and rational.

In the United States and other Western countries, engineers must be trained to grapple with the problems of integrating engineering with economic, financial, environmental, and social challenges. Few engineers are trained to consider these factors; unless engineers learn to deal with the broader issues, they may lose the status of an independent profession and become simple tools in the larger bureaucratic society.

The prospects of radical reductions in military spending world-wide, on which much engineering research and employment depended, may force engineers to redirect their interests and skill toward increasingly pressing economic and social needs. Engineers may need to redefine their role as well as responsibility in society and with it assume a more political and proactive approach.

EFFECTIVE ORGANIZATION OF ENGINEERING

To be effective, the modern organization of engineering activities requires interorganizational coordination. This coordination is achieved by a change from vertically integrated line or hierarchical organizations, which foster arms-length relationships and interpersonal competition, to flexible cooperative interorganizational relationships. A long-term commitment, investment by individuals in interpersonal relationships, and trust are the usual requirements for successful interorganizational coordination. It also demands openness, sharing of data, greater use of data information technology, and effective cross-boundary decision coordination.

Engineers can no longer be closeted in a narrow organizational slot with a defined design or operational responsibility. They must be encouraged to coordinate their work with others at various levels in the organization so as to assure timely exchange of information and incorporation of changes as the technology develops. Engineers do not perform well if hemmed in by narrow job definitions. Their creativity and even basic engineering performance will be affected by a lack of freedom to explore, develop, and create. Such confines usually stymie good engineers.

Effective engineering organizations are designed to solve problems and implement problem solutions effectively. They have clearly identifiable technological and often economic goals. Most engineering organizations are engaged in projects and are concerned with series of once-through engineering activities. They seldom engage in repetitive tasks.

Each engineering organization requires a leader. The role of such a leader is to inspire the people in what they are doing, guide them in their work, encourage them to do better, and support them in their performance and as individuals. A leader recognizes people for themselves, for their qualities as distinct individuals, and for their contributions to the objectives. Leaders learn to interpret broad objectives in terms of the values and interests of individuals.

ENGINEERING ASSOCIATIONS AND PROFESSIONAL BODIES

Engineers, like other professionals, are organized in various engineering associations, though in most Western nations these associations are defined by narrow engineering disciplines. In the United States, for example, there are nine major and no less than twenty-seven minor engineering associations, with no formal general engineering organization tying these associations together and representing engineers as a profession. The situation is not very different in Great Britain and the rest of the West, with the possible exception of South American and some southern European countries. Engineering associations concern themselves principally with technical matters and provide forums for the exchange and presentation of technical findings and information. They also engage in technical standard setting and provide venues for social gatherings and engineering product marketing. They scrupulously refrain from involvement in working terms, conditions, and standards of engineers, and they do not do much, if anything, to enhance or advance the standing and prestige of engineers—how different this is from the operations of associations in the legal and medical professions, or even professions such as accounting or banking.

In the United States, and to a lesser extent in other Western countries, the medical and legal professions are organized in powerful nationwide organizations that jealously guard and often control entry into their professions. They consider the maintenance of professional standards, working terms, and conditions, as well as prestige and status, their principal function, although most also provide their members with forums for the exchange of professional information and findings.

Traditionally, legal and medical professionals have considered mutual support and the maintenance of the status of their profession as their principal obligations. It is curious to note, for example, how seldom medical or legal professionals criticize one another even in the face of overwhelming evidence of professional misconduct or error. This is quite different in the engineering profession where engineers feel little professional responsibility for their colleagues. Engineers not only compete in business and as professionals, but they often will volunteer criticism of judgment on other members of their profession, without any objections by their associations or peers. Engineering peers often feel it to be their responsibility and democratic challenge to review publicly the performance of other engineers.

Engineering associations have never been able to define the term *engineer* and develop a specification of qualifications for the title. As a result, the public is confused not only in its understanding of what an engineer is but also with regard to what an engineer does.

7

Engineering and
Technology Development

Most technology has its base in scientific knowledge which is transformed by the application of experience to applied knowledge and then by the use of engineering to knowledge of how to design, produce, and use the evolving technology. Engineering, as shown in Figure 4, links scientific knowledge and experience to the evolution, development, and application of technology. Similarly, engineering plays an ever-changing role in technology development, as shown in Figure 5. After an invention is made by application of science and engineering, it evolves into a technology through innovation where experience and increasingly more detailed engineering develops improvements affecting the usefulness and applicability of the technology.

Engineers, to be effective in technology development, must have a solid understanding of scientific principles but comprehend that their primary role is that of an integrator and developer who translates scientific discoveries and experience into knowledge of how to develop, apply, and use technology based on the discoveries.

Engineers must do their jobs in an environment increasingly more conscious of and concerned with the economic, social, and

Figure 4
Engineering and Technology Development

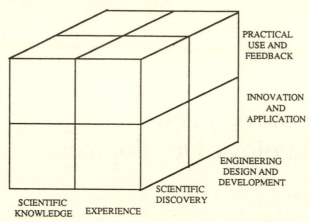

PRACTICAL
USE AND
FEEDBACK

INNOVATION
AND
APPLICATION

ENGINEERING
DESIGN AND
DEVELOPMENT

SCIENTIFIC
DISCOVERY

SCIENTIFIC
KNOWLEDGE   EXPERIENCE

environmental impacts of technology. In addition to a scientific base, they must have knowledge of the processes that serve to transform basic scientific knowledge into products, and they must know how to design the knowledge into something that can be produced and used to advantage.

This multiple and integrative function is expected to be the principal activity of engineers in the future as scientific knowledge advances more rapidly. To solve escalating problems of humankind, engineers will have to learn how to apply scientific knowledge and how to use the results of the development of that knowledge.

Although technology often is defined as the application of knowledge—which, according to Smith (1986), "is scientifically or otherwise derived and used to the creation or modification of things and processes"—it is useful to consider as technology all products, processes, and services that have or do contribute to some use, that offer some benefits, or that are otherwise desirable.

Engineering for many years was based firmly on developing useful, applicable technology and was concerned primarily with the transformation of scientific knowledge into new product, process, and service technologies. But engineering has become a more

Figure 5
Technology Development Cycle

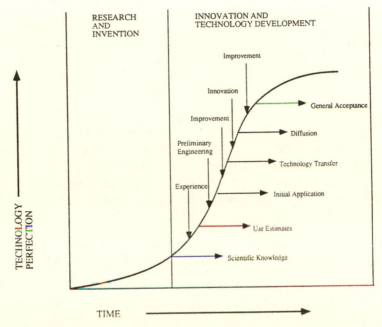

and more science-based and narrowly focused discipline in recent years, with much of the integrative characteristics required for effective technology development.

Engineers are trained to be narrowly focused on technology and to perform a specific engineering function, usually the design of a product or process. They expect to have a well-defined role and assume a specific responsibility in terms of the tasks they will perform; they do not expect to have an understanding of and concern for the broader issues involved or the implications of their work. A structural ship engineer, for example, will consider it his or her responsibility to design a most efficient ship structure, in terms of ship structural weight, which will perform the structural requirements imposed. He or she will not be concerned with the construction cost of the ship, its producibility, its maintainability, and how the structure performs as part of the ship. In other words, he or she will consider his or her role strictly functional and not interdependent or integrative. The most serious problem we face in technology development today is the functional specialization of engineers and the lack of understanding by many engineers that functional integration of all technical aspects and consideration of nontechnical factors and concerns are necessary for effective technology development.

In technology development, engineers are concerned with performance of their task and the progress of technology development. They are not involved in knowledge transfer from others and to others. This circumstance is unfortunate; only by effective networking of knowledge, experience gained, and assumptions or constraints used can continued progress toward improvements in technology be assured.

ENGINEERING AND PRODUCT DEVELOPMENT

Engineering capability is often questioned now as Americans and others compare the design of U.S. products with that of comparable or identical products engineered abroad. While discrepancies sometimes are noted in terms of product quality and performance, many products engineered in the United States and other Western countries are uncompetitive because they are not engineered for efficient production, assembly, use, and maintenance. This problem exists not because of a lack of technology or because of high material and labor costs, but simply because of bad or ineffective engineering and design.

The problem here is not a function of a shortage or the training of engineers. The United States has a larger number of engineers per capita than other major industrial nations, such as Germany and Japan. Similarly, U.S. engineers are on average equally well trained, yet U.S. training is more theoretical and U.S. engineers are not taught how to apply their engineering skills to the solution of real problems. They similarly are not trained to integrate design and manufacturing or to choose designs and materials with the product use and user in mind. Primary product costs are an over-riding consideration in the United States, and marketability is judged only from the manufacturer's cost viewpoint.

Engineers are told to design an adequate product at the least possible cost, and they do not get an opportunity to feel responsible for the product. They usually are part of an anonymous group that works as a functional organization without any proprietary relationship to the products under design.

To be effective, engineers must be organized around the product, use, and customer. Engineering must be done close to the product and process, and engineers must be involved in all stages of product development, design, manufacturing process selection, and product use and marketing. Information should be fed back and forth between all the different stages from product conception to ultimate use.

Few U.S. engineers spend time on the shop floor in marketing or in the field. In fact, few have any conception how their product is manufactured, marketed, and used. As a result, engineers often lack knowledge about how to make a product more producible and usable.

Engineers usually design products and processes from scratch, even if similar products and processes exist or major components or assemblies that are needed for the product are readily available from suppliers. Engineers are not used to making design or buying tradeoffs. Few engineers have any idea how much things cost to buy or to make. They will often, as a result, try to minimize such things as steel weight in a ship hull, even when a 1 percent savings in steel weight adds 10 percent to construction costs and, furthermore, makes the resulting ship less usable or maintainable. Engineers usually do not know the economic value of a steel weight saving.

It is not enough to catch up in manufacturing technology and to maintain a dominant research and development base. Unless a country is effective in engineering, which provides the glue that joins research and discovery to product manufacture and use, most of the inventiveness and basic technological advances are wasted because they will not be developed to an effective use in a timely manner.

8

Engineering Education

Engineering education in the United States has grown out of land grant college curricula, which were designed to train practical problem solvers who would be able to improve the agricultural productivity of the country. This approach has served the United States well and, though agricultural engineering has declined in university programs, the agricultural sector continues to benefit from past advances, which even now make U.S. agriculture the most productive in the world.

Engineering for manufacturing and industry had quite a different history. During the nineteenth century, inventors and entrepreneurs in the United States worked alone or for small enterprises. As a result, it would take nearly 100 years for engineering to develop as a formal area of study. By contrast, in Europe large industrial groupings emerged during the Industrial Revolution, which demanded large numbers of engineers. Engineering education emerged in Europe as a formal discipline late in the eighteenth century.

Engineering education has gone through a number of cycles in the United States. While oriented largely toward problem solving in its early years, it became more science-oriented and theoretical starting with World War II. Only in recent years has engineering education started to reemphasize problem solving and practical issues such as manufacturing.

Engineering education is largely concentrated on the interpretation of science in engineering terms. It consists largely of rigorous theoretical training, with little encouragement for technological or scientific creativity. More important, it discourages applied studies. In other words, today's engineering education develops an engineering culture that is rigorous in the narrow sense of technology. It also isolates engineers from the real needs or wants of society. In fact, it encourages engineering for engineering's sake. Engineers also are not trained to be involved in decision or policymaking. Engineering education by and large contributes little to engineers' ability to play a leading role in guiding technology development and use.

In Europe technical institutes called *polytechnics* were established in the late eighteenth century; such institutions emerged in the United States about 100 years later. The sciences, particularly the natural sciences, began to flourish as formal disciplines early in the nineteenth century and started to impact on engineering education as the link between science and engineering. Until that time engineering knowledge was largely derived from experimentation or other types of observation. Engineering thus became the bridge that linked scientific discovery and knowledge to practical application via technology.

Engineering education has since evolved into a more science-based approach, mostly concerned with teaching principles and fundamentals. As a result, engineers learn to understand what makes technology work but gain no knowledge of the potential use or function of technology. They are educated in understanding the scientific base required for the creation of new technology. They learn little about the role of technology in social, economic, and industrial development which is needed to generate new products and processes to meet the needs of society.

Engineering education is concerned largely with the teaching of narrow principles and often discourages consideration of the larger issues and problems, which are assumed to be the responsibility of others who by intuition or magic are able to define the narrow technological opportunities or needs which the engineer is then supposed to address by developing an engineering solution.

PREPARATION FOR ENGINEERING EDUCATION

Although the United States spends on average more on education than any other industrial country, educational performance continues to decline. The average annual cost per high school student now exceeds $6,000 per year, with some school districts spending as much as $7,880 per year. At the same time, SAT (Standardized Achievement Test) scores for college admission fell to 896 out of 1600, with verbal scores now at a dismal 422 as shown in Figure 6. After inflation, funding for U.S. schools increased by 30 percent since 1980, or by a compounded rate of over 7 percent per year. Teachers' salaries have nearly doubled since 1975, but administrative costs and costs for extracurricular activities have increased even more rapidly. The U.S. school year is among the shortest of any developed country. This is compounded by the fact that the average school day is also 16.2 percent shorter than that in Japan and 10 percent shorter than that in Germany. Furthermore, more time is spent on extracurricular or soft subjects. As a result, the direct class hours devoted to communication, mathematics, science, and similar basic areas are only about 60 percent of those devoted to these subjects in other industrialized countries. It is no wonder that we graduate many functionally and scientifically illiterate people at a time when technology puts an increasing demand on people in the workplace.

Higher education in Western countries is concerned largely with science; the social sciences; specialized professions, like law and medicine; and general management. The number of science graduates in Western countries is larger than that of engineers at both the undergraduate and graduate level, and both science and engi-

Figure 6
SAT Scores

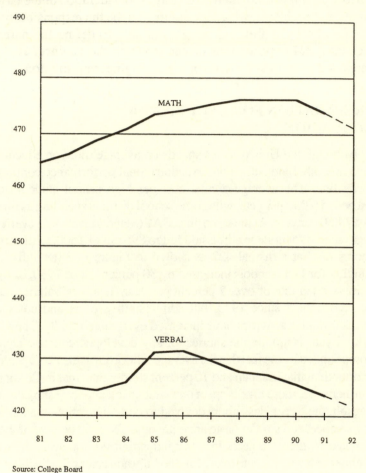

Source: College Board

neering graduates are far outnumbered by social science, humanities, general management, and similar graduates.

The United States is unusual in that the majority of high school
graduates attend college, though most enter liberal arts programs,
with little or no professional objective in mind. The large number
of college students is in part facilitated by government and private

financial aid awarded without regard to program or field of study selected. Financial aid for both undergraduate and graduate education leading to professional fields such as science, engineering, law, and medicine generally has been less generous than for liberal arts programs. Apparently, donors believe that students in fields with a potential of large earnings need less financial aid in their education.

In the United States undergraduates often change major fields in response to changes in the job market or perceived status of a field. Most shifts of science or engineering students, however, were, until recently, within these fields, though an increasing number are now opting into management, law, or medicine.

Less than 20 percent of U.S. baccalaureates are awarded in the natural sciences and engineering, of which 42 percent are in engineering. Yet even these numbers are misleading, as nearly half the engineering baccalaureates and over 88 percent of the graduate degrees in engineering are in engineering science and not engineering.

Figure 7 shows the flow of U.S. students into and out of natural sciences and engineering education between high school and graduate school. It indicates that of high school seniors who choose natural sciences or engineering, only 22 percent actually obtain a bachelor's degree and only 6.3 percent obtain a graduate degree.

Undergraduate participation in research, quality teaching, and supportive, individual attention to student development are essential to encourage students to continue in science or engineering. Such a research-oriented environment is found in only a few undergraduate institutions. At the graduate level, Ph.D.-level education is concentrated in about 100 of the 330 universities that award science and engineering Ph.D.s, yet only thirty-eight institutions award doctorates in engineering.

In Japan, on the other hand, engineering graduates outnumber science graduates by a factor of over 5.8, and both together outnumber graduates in other professional disciplines, such as law, medicine, and economics. The number of degrees in the physical sciences and engineering in the United States and Japan in 1986 is shown in Table 1.

Figure 7
Student Flows into and out of the Natural Science/Engineering Pipeline:
The High School Senior Class of 1972

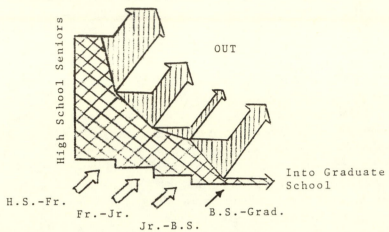

Source: Hilton and Lee (1988).

Therefore Japan graduates more than twice the number of engineers per capita than the United States. This ratio is significantly larger still when compared with the data for some Western European countries. Also, engineering in the United States and other Western nations generally is less respected than other professions. Medical doctors, scientists, lawyers, and bankers usually rank higher in public esteem than engineers. This fact may be due in part to the linguistic anomaly that lumps basic mechanics, building heat plant operators, and highly qualified design or research engineers under the one title of *engineer*. In Japan and other Far Eastern industrialized nations, engineering is the most respected profession.

Engineering in continental Europe, particularly in France and Germany, is largely approached as "practical science" and is taught at special engineering schools. A growing number of students are opting to become engineers as a result of the recognition that attends these special schools.

Table 1
Degrees in the Physical Sciences and Engineering in the United States
and Japan, 1986

	U.S.	Japan	Ratio
Physical Sciences			
Bachelor	83,859	11,803	7.1
Masters	15,318	1,710	9.0
Doctoral	7,374	822	9.0
Total	106,551	14,335	7.4
Engineering			
Bachelor	71,094	75,188	0.9
Master	18,550	6,975	2.7
Doctoral	2,742	1,186	2.3
Total	92,386	83,349	1.1

Source: Hilton and Lee (1988).

FEDERAL GOVERNMENT PROPOSALS

The Office of Technology Assessment of the U.S. Congress in its report brief on "Educating Scientists and Engineers" proposes two broad strategies for federal action to replenish and enhance the supply of scientists and engineers. The strategies complement each other and operate best in tandem. The first is a *retention* strategy. Its primary goal is to retain undergraduate and graduate students well along in the educational process by reducing attrition and, especially in the case of undergraduates, by encouraging them to go on for a Ph.D. Such short-term, proven retention efforts could modestly increase the supply of scientists and engineers within a few years. The second strategy is an attempt to enlarge the base of potential scientists and engineers over the long term by recruiting more and different students into science and engineering study. Such a *recruitment* strategy entails working with schools and colleges, and with children, teachers, and staff, to improve elementary and secondary mathematics and science education. Together these two strategies can raise the quality, broaden the demographic composition, and (if necessary) increase the number of young scientists and engineers to meet near-term research and development needs and prepare for longer-term, uncertain needs.

POLICY OPTIONS TO IMPROVE SCIENCE AND ENGINEERING EDUCATION

Recruitment—Enlarge the Pool

- Elementary and secondary teaching: encourage and reward teachers; expand support for preservice and inservice training.

- School opportunities: reproduce science-intensive schools; adjust course taking and curricula; review tracking; review testing.

- Intervention programs: increase interest in and readiness for science and engineering majors; transfer the lessons from successful programs; encourage sponsorship from all sources.

- Informal education: increase support of science centers, television programing, fairs, and camps.

- Opportunities for women: enforce Title IX of the Education Amendments of 1972; provide special support and intervention.

- Opportunities for minorities: enforce civil rights legislation; provide special support and intervention.

Retention—Keep Students in the Pool

- Graduate training support: "buy" Ph.D.s with fellowships and traineeships; these people are most likely to join the research work force.

- Academic R&D spending: bolster demand and support research assistants, especially through the mission agencies.

- Foreign students: adjust immigration policy to ease entry and retention.

- Undergraduate environments: support institutions that reward teaching and provide role models, such as research colleges and universities and historically black institutions.

- Hands-on experience: encourage undergraduate research apprenticeships and cooperative education that impart career skills.

- Targeted support for undergraduates: link need- or merit-based aid to college major.

Strengthen Federal Science and Engineering Education Efforts

- National Science Foundation as lead science education agency: underscore responsibility through the Science and Engineering Education Directorate for elementary through undergraduate science programs.

- Federal interagency coordination and data collection: raise the visibility of science education and the transfer of information between agencies and to educational communities.

OUTPUT OF U.S. ENGINEERING EDUCATION

U.S. universities continue the trend of educating a slowly growing number of engineers, largely trained for engineering research or engineering design. Few are trained as technologists who solve engineering problems or improve the application and use of technology.

Since 1967, Japan has consistently graduated more engineers with bachelor's degrees than the United States, as shown in Figure 8. In Japan nearly all engineering degrees are granted to Japanese students; in the United States over 10 percent of the engineering baccalaureates and nearly 23 percent of graduate engineering degrees awarded are granted to foreign nationals. In Japan the baccalaureate is the terminal professional engineering degree, with more advanced engineering training carried out by employers; in the United States nearly 58 percent of engineering graduates continue toward an advanced degree, though

Figure 8
Bachelor's Degrees in Engineering in the United States and Japan

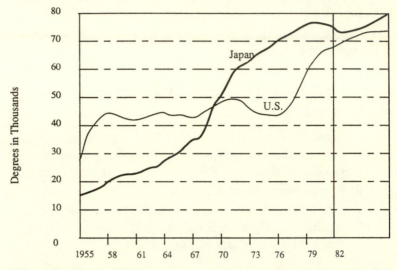

Source: Hilton and Lee (1988).

an increasing percentage pursues graduate study in other fields such as management, medicine, or law.

As a result, engineers entering the U.S. job market with either a bachelor's or master's degree in engineering is now only about 48,000 per year after reducing the total number by

  1. those entering other fields,

  2. foreign citizen graduates who do not seek U.S. employment, and

  3. engineers who continue toward a research career.

U.S. industry today obtains an inflow of only 60 percent as many newly graduated engineers as Japan. On a per capita basis, this translates into only 30.2 percent as many engineers, a percentage that drops even further when compared in economic terms, or in terms of the number of engineers per unit of national output.

An additional dilemma is the fact that the U.S. government and other nonindustrial organizations absorb about one-quarter of all U.S. citizen engineering graduates (compared to about 10% in Japan); ultimately, U.S. industry has to make do with an influx of graduate engineers that is only about 22 percent of that obtained by Japanese industry on a per capita basis and about 19.6 percent in economic terms.

As noted before, the United States awards many more graduate engineering degrees that Japan and other countries. In recent years, the number of master's degrees awarded in engineering was about 38 percent of the total number of degrees awarded by U.S. universities in engineering, while doctoral degrees in engineering averaged about 7 percent of total engineering degrees awarded in the United States.

ENGINEERS IN RESEARCH

Engineering and science have moved closer together, and engineering research often now involves scientific investigations. Yet one quality that facilitates scientific discovery, and is a sine qua non in scientific research, particularly in experimental scientific research, is the mental ability to play around, to consider the illogical, to cultivate some chaos, and to observe or analyze the unexpected. Successful scientists consider results as significant even when they are not what scientists are looking for or when they were obtained by chance or as a side effect. By contrast, engineers are methodological, rational, and analytical; they walk a planned defined path and abhor surprises such as unexpected results. Not only are they risk averse, but they are trained to solve problems by a step-by-step approach. Scientific discovery, though never entirely accidental, holds elements of surprise. Engineers have difficulty coping with such uncertain, chance developments. Most engineering research, as a result, is undertaken with foreseeable results rather than with results that pose a likely surprise.

ENGINEERING RESEARCH AND EDUCATION

Engineering research and development deals largely with basic science problems. This focus has resulted in major changes in graduate engineering programs which are geared more at training research assistants, future researchers, and faculty than professional engineers. Many academic institutions siphon off the best and brightest of their engineering students, with relatively few left over for industry.

Industry increasingly is disappointed with the effectiveness of conversion of basic academic research to technological innovations and business opportunities; as a result, it has lost much of its interest in supporting engineering R&D at universities. A general belief prevails that academics have little, if any, concept of the problems of society and how engineering research contributes to their resolution. Interest in the application of scientific-engineering research to real-world problems is considered too mundane and too applied.

At the same time, academic researchers have lost much of their interest in industrial problems and generally look down upon applied engineering research as intellectually inferior. This situation is due in part to the lack of effective communication between industry and academia, but it also is affected by the fact that fewer and fewer academic faculty and researchers have any industrial experience. The vast majority of U.S. engineering faculty members have spent all their professional lives at a university, and often the same university.

Similarly, research agendas usually are devised by disciplinary peers that perpetuate the inbreeding system. Most engineering research at U.S. universities is performed at institutions that have strong graduate programs oriented largely toward the training of research assistants, future researchers, and research faculty, but not engineers.

Graduate, and to some degree undergraduate, teaching is designed to support research activities; research activities are not designed to support teaching, although research obviously contributes to the quality of at least graduate teaching. The link, however, has become more and more tenuous. An increasing number of graduate subjects are developed on the basis of new research findings. While this approach is admirable for programs designed to train future researchers, the effectiveness of this approach in advancing engineering education is not so obvious.

The main problem to date is the increasing gap between researchers and practicing professional engineers. The research engineer selects a research topic and after approval designs the research and performs the investigation of the research. Successful completion of the research consists of achieving the research objectives, which are often clouded in general principles with little, if any, thought given to the contribution of the research to real-life problems. Research engineers seldom are trained to understand how the research results can be used to further development, innovation, and implementation followed by use or utilization of the technology.

As engineering graduate education becomes largely a breeding ground for science and engineering researchers, professional engineering education has lagged. Unfortunately, the application of science, the principal function of engineering, is not as respected as doing science. As a result, we increasingly find graduate engineers educated in doing and not applying science.

Consequently, we see a decline in the growth and quality of graduate engineers who go into engineering. The number of engineers and scientists in R&D in the United States grew from 450,000 in 1965 to nearly double that number in 1988 (see Table 2). During the same period the number of practicing professional engineers declined from over 688,000 to just over 650,000. In Japan the number of scientists and engineers in R&D nearly quadrupled during the same period, yet the number of professional engineers increased more than twofold.

In the United States the number of scientists and engineers working in nondefense R&D was 316,000 (70.2%) in 1965 and grew to 552,000 (66.8%) in 1988, while in Japan 174,000 (97.7%) and 389,900 (95.8%) of scientists/engineers worked in nondefense R&D in 1970 and 1988, respectively. These differences are re-

Table 2
Scientists and Engineers in Research and Development (in thousands)

	USA	JAPAN	GERMANY	FRANCE
1965	450	98	52	46
1970	495	178	80	49
1975	508	279	96	54
1980	592	310	112	58
1985	720	362	116	68
1988	832	408	119	72

flected in the ratio of expenses for nondefense R&D as a percentage of gross national product (GNP) in the United States. This ratio grew from a paltry 1.6 percent in 1970 to a still low 1.8 percent in 1988, while Japan achieved growth from 1.8 percent in 1970 to over 2.8 percent in 1988.

While the United States has spent nearly 2.8 percent of GNP on total R&D expenses throughout this period, only 54 to 66 percent of that amount was spent on nondefense R&D. By contrast, Japan and the former West Germany spent nearly 1 percent more of GNP on nondefense R&D throughout much of the same period.

RELEVANT ENGINEERING RESEARCH

Society increasingly expects engineers to perform relevant research directed toward the solution of real problems. Unlike scientists, engineers are assumed to deal not only with the extension of knowledge but with the improvement of societal conditions. The public looks to engineers for the improvement in technology with clear economic or social goals.

Scientists are expected to do research solely for the sake of knowledge, without any preconceived goals and without focusing on a particular problem. But engineers are supposed to address problems and issues of concern. Engineering education must prepare engineers to be effective problem solvers. Engineers must learn to recognize problems and their implications and must be trained to develop meaningful approaches for their solutions.

Engineers and engineering research should have clearly identifiable technological, economic, and societal goals. Engineering research, by its very name, should be concerned more with technology development than with scientific discovery and principles. It should build upon these discoveries and principles and develop means for the application of them to potentially useful technology. It should be an innovation process that transforms initially vague scientific results into real solutions. Engineering research should not end with the transformation of science into useful technology; it should be responsible for the continued improvement of the technology until it approaches maturity and ultimately obsolescence.

We in the United States are not good at this process. The overemphasis of engineering science has encouraged most engineering research to be devoted largely to the development or improvement of scientific principles, with comparatively little effort directed at the transformation of these principles into the development of technology. Even less effort is devoted to the continuous development and improvement of technology. The Japanese, by contrast, have been particularly strong in such endeavors.

While we may not want to overemphasize applied engineering or technology innovation research, we certainly need a better balance. Most important, we must make it more respectable to assure that our best engineers do not concentrate mainly on scientific engineering research. We have an abundance of inventions and scientific discoveries. Much of our knowledge base is unused. We lack adequate engineering involvement in technology development and innovation. Unless we learn how to reorient our priorities, the motives, and rewards in engineering research, we probably will continue to lose our technological leadership.

JAPANESE UNIVERSITY EDUCATION

The Japanese university system works as a stifling, restrictive, and self-righteous hierarchy that generally discourages independent thought and unconventional approaches in teaching and research. Junior faculty and other researchers, which means anyone under the rank of the professor of the department or the chair, are expected to toe the line and be yes men. Universities recruit faculty mostly from among their own graduates and promote them by seniority only. Japanese universities are more exclusive or "clubby" than their European or U.S. counterparts.

Faculty careers are predetermined and are concentrated on the emphasis and research interest of the chair or professor. Junior faculty members are not expected to, and are discouraged from, carrying out independent research or writing. They are expected to help with teaching and with the research and writing of the professor. They are expected to work their way up through the *koza* as graduate assistants and qualify for lifetime appointments, but their advancement usually is determined by strict seniority. Most university research is funded by the Ministry of Education, which distributes money according to its interpretation of the needs of the state for graduating engineers and for advances in research.

Few professors in Japan run or participate in business, and consulting is not encouraged, at least at the six national universities where professors are civil servants subject to civil service payscales. As a result, more and more basic, as well as applied, research is done by industry in industry-funded research institutions, which increasingly attract young engineers and researchers who want to use their own ideas.

Industry provides most graduate education either internally or by sponsoring engineers at foreign universities. Industry is developing large-scale in-house training schemes that are often a combination of advanced study and applications, including rotation among company departments.

There is today a small movement aimed at changing the traditional structure of Japanese universities and making graduate study and research more accessible, more rewarding, and more application focused.

NEW DIRECTIONS IN ENGINEERING EDUCATION IN THE UNITED STATES

Engineering education rightly has focused on teaching scientific and mathematical principles in the early undergraduate years. Later undergraduate years and graduate education are devoted largely to the expansion of these principles. When applications are introduced, they are introduced as theoretical concepts selected for their example in the use of the theories rather than for their engineering reality. As a result, few engineering students learn how to define, structure, and solve real-world engineering problems. The reason for this shortcoming appears to be the result of inbreeding of engineering faculty, few of whom have had any contact or experience with real-world engineering. To them, engineering problems are interesting theoretical engineering issues that apply a particular theory.

Engineering education, particularly in the United States, requires a radical change. It is not that leading U.S. institutions do not deliver a high-class science and engineering education, but that they do not train engineers to understand the environment in which technology performs and the broader issues that technology addresses and affects. Engineering students should learn to identify problems that require technology in their resolution; they should not just invent new scientific principles or technologies to have someone else search for a possible application that often results in misuse.

Students should be encouraged to consider even simple engineering problems in context and be allowed to make up their own problems to test their skills in applying theory to problem solving. They should be encouraged to think and to puzzle problems out using unbridled creativity as an aid, even if it does not result in the most "efficient" approach to a problem solution.

Such an approach not only would teach a student to think for himself but also would train him or her to identify issues and solution approaches independently. Such skills are necessary for successful engineering in the real world.

Engineers should learn to consider product and process development, economics, and environmental impacts from a systems point of view, not only from a design point of view. To be effective, engineering education must tie together science, engineering, management, and society. Engineers should become aware of the following:

- Science generates technological advance.
- To be effective, technology must be engineered not only in terms of the scientific principles involved but with full consideration of user and societal requirements.
- The development of technology demands effective management.
- The use of technology needs well-planned and executed management to assure societal and environmental acceptability of new technology.

Engineers must be educated to understand the processes that link science and technology, the engineering activities that permit scientific principles to evolve into useful technology, the processes required to implement the technology, the uses and conditions under which the technology is applied, and the societal and environmental impacts of the technology. The various interactions among science, engineering, and technology development are shown in Figure 9.

Engineers must learn how to deal with all the processes as well as the human activities involved in the link between scientific discovery and the ultimate use of technology. As the world becomes more and more dependent on science and technology, not only for economic growth but ultimately for survival; as population grows; as resources become scarce; as the environment becomes more polluted; and as society becomes more demanding and complex, engineers must prepare for and accept a greater role in decision making. The ability to make the needed decisions and judgments demands education in science, engineering, technology,

Figure 9
The Linkages between Society and Technology

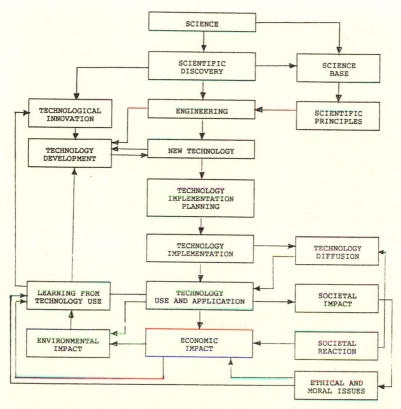

social and human relations, communications, management, the environment, and ethics.

Most of these decisions, even though they involve technological knowledge and judgment, are made today by leaders with no scientific or engineering background. Some may use expert input to help in the decision making, but ultimately the nontechnical decision maker makes the choice. Few senior executives in Western industry have a scientific or technological background, and in the military fewer than 5 percent of flag officers (generals and admirals) have a technical education. This situation must change if we do not want to put our economic, societal, and environmental future at risk.

NEED AND ROLE OF GRADUATE ENGINEERING EDUCATION

Medicine, law, management, and other professional disciplines in the United States have been based firmly on graduate-level education. Engineering has been one of the holdouts. A bachelor's degree in engineering is still considered by many the entry-level qualification for engineers in industry and government. Advanced degrees, while useful, are largely thought of as requirements for careers in research or teaching.

Engineering schools often start to focus on narrow engineering disciplinary requirements during the third and fourth year of undergraduate study. For engineers continuing with graduate study, the focus becomes increasingly narrower and leads toward highly specialized scientific engineering or research concentrations. Some claim that graduate programs in some engineering schools aim more at training researchers than graduate engineers.

With a large percentage of graduate engineering students working as graduate research assistants, narrow disciplinary research-oriented focusing starts early during initial graduate study and becomes increasingly focused if study is continued toward an advanced degree, such as a doctorate. The percentage of graduate engineering students working as research assistants is many times that of law, management, architecture, or medical students in such jobs.

Students in law, medicine, and so on do not serve as research assistants during their early years of graduate study because their studies do not deal with narrow research issues. Only after completing their professional education will students in these disciplines start to engage in focused study and research, and then only if they are interested in advancing and not just applying the knowledge they have acquired. If they are not research oriented, they will try to gain professional experience as a clerk to a justice, an intern in a hospital, and so on.

Only a small proportion of law, medical, management, and similar graduate students engage in narrowly focused research for any length of time; if they do, it is only because they are interested in research or a research career. In engineering, by contrast, many graduate students serve as research assistants from the start of their graduate study. The large salaries and earnings expected by law, medical, or management graduates may explain their willingness and ability to finance graduate professional study without working as research assistants while in school. Engineering counterparts may not have this luxury.

Because engineers expect to earn less, they may be less willing to finance (or borrow for) their graduate education. But, it also may be that because their graduate education is not oriented toward broad professional training, they are offered only lower-paying specialist positions.

Recently it was suggested that engineering students be required to spend 5 years in college before receiving a first engineering degree, say a joint B.S./M.S. in engineering; many objected not because they disagreed with the assumption that it takes 5 years to educate a professional engineer but because they felt that few prospective engineers would be willing or able to finance 5 years of engineering study. In a way, this is an admission that a professional engineering degree is not worth as much of an investment as a law, medical, or management degree for which students willingly go into debt.

It is not clear whether holders of engineering degrees that clearly are oriented toward comprehensive professional engineering would be offered significantly higher salaries and earnings. Many in industry believe that engineers' remuneration would be well ahead that offered engineers today if professional engineering education really was oriented toward the broader needs and provided a new generation of engineering leaders and managers.

Many, including the author, believe that engineering education in the future should require not just an extended bachelor's program of 5 years leading to a joint B.S./M.S. degree in engineering, but that it should consist of a first or pre-engineering degree at the bachelor's level, after 4 years of study, followed by a graduate engineer's degree after another 2 years of graduate study. Such a degree could be called a master of engineering, a master of engineering management, or a master of engineering science, to distinguish the three special areas of emphasis offered to graduate engineers.

SUGGESTIONS FOR A NEW ENGINEERING CURRICULUM

To satisfy the needs of society and industry in the future, engineers not only will have to become proficient in engineering sciences but will have to acquire a broader education and the broader skills of leaders, managers, and decision makers.

The first 4 years, or the pre-engineering education, may consist of 2 years of study of fundamentals in the following:

- Mathematics—4 terms
- Physics, thermodynamics, mechanics, dynamics, and so on—4 terms
- Materials—2 terms
- Chemistry—1–2 terms
- English and literature—2 terms
- History and geography—2 terms
- Interpersonal relations—1–2 terms
- Art or music—1–2 terms
- Computer technology and programming—1–2 terms
- Other—1–2 terms

The next 2 years would focus on engineering fundamentals that are built on the mathematics and science foundation established in the first two years. The third and fourth year should introduce the student to the following:

- Applied mathematics

- Engineering design
- Engineering analysis
- Manufacturing and operating principles
- Technology operations management
- Computing, computer control, computer-aided design and manufacturing
- Engineering management
- Statistics and quality management
- Interpersonal communications
- Project management
- Introductory courses in focused areas of mechanical, civil, electrical, aeronautical, or other areas of engineering
- Organization behavior and leadership
- Financial management
- Other

On completion of these 4 years, the student will have qualified in pre-engineering and will be ready to start his or her 2-year professional engineering education. Here the student normally would be expected to choose not only an area of engineering but also a concentration such as engineering (design and operations), engineering management, or engineering science. Any combination of area and concentration would lead to a professional engineering degree. Students may opt to continue to an advanced degree, say a Ph.D. in engineering.

All professional degrees in engineering would include required courses in the management of technology, management of engineering, leadership and organization, and project management. These courses usually would offer students an opportunity to relate the material of the course to engineering design and operations (including product and process development), to industrial management, or to the management of R&D.

It is suggested that graduate professional engineering education include engineering internships of at least 2 graduate terms (e.g., two summers). Internship would be served in industry and should be divided into one-half terms, one each in engineering/R&D and in manufacturing/marketing. Engineers in all three concentrations—engineering, engineering management, and engineering science—should serve internships in the four major activities of industrial firms to assure their familiarity with real-world conditions and requirements. Similarly, graduate engineering students would select internships in their engineering area.

The idea is not only to educate engineers in science and engineering but to give them an understanding of the environment in and the methods by which technology is developed and used, as well as to equip them with an ability to communicate, lead, and manage in our increasingly more complex technological and social environments.

9

Engineering as a Profession

Engineering not only is undefined in the public mind but is quite fragmented in actual practice. The qualifications required to practice engineering vary widely, with no qualifications at all required in some locations. One reason for this fragmentation is that few authorities have ever defined what an engineer is.

There are well over 100 different engineering specialties from the basic mechanical, electrical, civil, aeronautical engineers to much more narrowly defined engineering disciplines. Many of these disciplines overlap widely, while others are rather ill defined. Electrical engineers, for example, have an ocean technology division, while civil engineers, marine engineers and naval architects, marine technologists, and many more claim ocean-related engineering and technology their domain of interest. Engineers not only are diffused professionally but the function of their professional associations is usually quite narrow. Their associations typically serve as forums for discussion of technical issues, as social centers of like-minded individuals, and sometimes as a conduit for the dissemination of technical and commercial information. They typically hold meetings, sometimes with exhibitions and technical sessions, and publish technical journals.

Engineering associations do not provide professional guidance, establish professional standards, lobby for the status and other benefits of the profession, or award professional credentials. They similarly play little, if any, role in overseeing licensing arrangements.

REGISTRATION AS A PROFESSIONAL ENGINEER

All fifty-five states and/or territories of the United States have enacted individual legislation to safeguard life, health, and property and to promote the public welfare. Registration of engineers under this legislation is a voluntary process and is generally achieved by compliance with three requirements: education, satisfactory experience, and examination. If an examination is given it usually covers the fundamentals of engineering (FE) and the principles and practice of engineering (PE). Four or more years of qualifying work experience acceptable to a state board are necessary to be eligible to take the PE examination provided all other statutory requirements are satisfied. The requirements for licensure vary among the states; however, once licensed, an engineer's registration will be recognized reciprocally by other jurisdictions upon submission of the appropriate verified documentation and satisfaction of other prerequisites.

Engineers in private practice usually must be registered, depending on their level of responsibility. Registration requirements vary somewhat in each of the states, and boards of registration for professional engineers determine if and when engineers are eligible to take the FE exam. All states use the National Council of Engineering Examiners (NCEE) FE exam as the first step in the registration procedure, although subsequent steps vary widely among states.

According to a recent survey, less than half of all U.S. engineering students take the first part of the professional engineers' examination prior to or immediately after graduation, and only half of these complete their license examination. Only 13 percent of engineering schools actually require their students to take the examination and only 3 percent require them to pass.

Many feel that engineers in the United States should take the idea of professional licensing more seriously. Becoming a professional engineer today should include obtaining a government-sanctioned and -issued license. Membership in voluntary professional societies is no substitute for licensing. It is unfortunate that engineers at large have such a low opinion of the value of licensing. Their opinion may be due to the low level of qualifications and knowledge required to pass licensing requirements, but it is probably the result of the lack of stature of and need for such a license.

Many large engineering firms have a few "licensed" staff members who sign off designs when needed, even when they are not directly involved in the engineering themselves. Engineers must recognize that their disparaging attitude toward licensing and their refusal to make a license a requirement for the practice of engineering only hurts engineers as professionals. Their attitude essentially indicates that engineers themselves feel that no special threshold of qualifications is required to practice engineering. If the level of license requirements is too low, then let us change that. But change will occur only if senior and respected engineers get involved in redesigning the licensing process and then require all engineers to hold a license. Such change appears to be an essential step toward transforming engineering into a true profession, with both responsibilities and rights, and assuring that engineers meet the needs of the U.S. economy and industry rather than the needs of the research community.

ENGINEERING LIABILITY

In recent years liability has assumed crisis proportions. Although liability of medical doctors tops the scale in terms of professional liability awards both in numbers and monetary value, engineers increasingly are being subjected to liability claims as well. Although government regulators traditionally assume the responsibility of protecting the public from harmful products, processes, and services—including infrastructure and buildings— courts increasingly assume the role of technological protectors and thereby become judges of technology.

At a recent meeting of the U.S. National Academy of Engineering, Peter Huber, a lawyer and engineer, noted that courts view technology and what engineers do in a "reactive sense and consider new products, processes or procedures to be guilty until proven innocent." In other words, engineers who use new approaches, nontraditional (or unproven) designs or concepts are more likely to be held liable if something goes wrong than those using or repeating traditional, so-called proven, approaches. The threat of liability obviously discourages engineers from the use of new innovative technology and approaches, no matter how effective, economical, or even socially and environmentally attractive (see Appendix D).

Engineers feel comfortable in this environment, particularly as they, unlike other professionals, do not feel any congenial professional responsibility for each other. They are quite willing to appear as expert critics (or witnesses) against other engineers.

Engineering liability not only has grown in terms of the number of cases filed and awards made; it also has become increasingly adversarial. The inability of lawyers and judges to make effective determinations of cause and effect and therefore fault, compounds the problem.

Liability costs introduce added costs to the design, manufacture, and use of technology. Unless conditions change radically, technology development increasingly will become defensive. Such a development would put the United States at an additional competitive disadvantage in relation to technology developments elsewhere, particularly in the Far East (see Appendix E).

INTERPERSONAL COMMUNICATION AND BEHAVIOR AMONG ENGINEERS

One of the most interesting things about engineers is how they communicate and relate with each other. Engineers easily develop narrowly specialized semantics which makes their communication somewhat exclusive to a narrow group of engineering experts. In fact, each little engineering fraternity has its own lingo and tenaciously holds onto what it considers its professional language. Engineers may consider this system a sign of exclusivity, but in reality it does nothing but stymie professional communication, particularly as other engineers in related fields often use very different terms for the same object or issue.

Engineers are even worse in communicating with nonengineers and often have great difficulty translating simple engineering concepts into laypersons' language. This situation is unfortunate as engineers are assumed to translate and transform science and scientific principles to practical applications.

Many engineers not only have difficulty in communicating verbally but also find it difficult to express their thoughts clearly in writing. Engineering reports, with few exceptions, are literary disasters. They often lack formal sentence structure and abound in poor grammar. This problem may be due to the fact that engineers have little time and patience to "put things into words" but more likely is due to their inability to express themselves properly.

Another issue that affects communication by and among engineers is their lack of constraint criticizing and evaluating the work and/or opinion of other engineers. Professionals such as lawyers and doctors, for example, make it an unwritten but accepted rule not to wash dirty linen in public. They do not criticize or evaluate other colleagues in public or in discussion with nonmembers of their profession. Even among themselves, they usually refrain from outright verbal criticism.

How differently engineers behave. They seem to thrive on putting other engineers down and to criticize their work or opinion particularly in the company of nonengineers. Engineers take great pleasure in explaining why a colleague's approach or suggestion does not make sense or is unsound or nonscientific—the harshest of all criticisms. This zest for criticism is unfortunate, because it really tears the engineering fraternity apart and makes it all the more difficult to get some coordination among engineers toward raising the status and remuneration of engineers.

Lack of communication skills also may affect the role of engineers among peer groups at the working or management level, as engineers often lack the personality for or approach to effective group leadership. Such leadership requires interpersonal and communication skills, an understanding of people, and a knowledge of how to get along with people. Leaders are sensitive to the feelings of others and know what to say under various conditions and how to relate to different people.

Leadership skill includes effective speaking, attentive listening, sensitive interaction, diplomatic interpersonal relations, tact, and compassion under appropriate circumstances. Leaders also use criticism carefully, particularly when those at fault have tried their best or know that they have not done a good job. Encouragement and positive valuation of the opportunities are much more effective in moving others toward better performance without self-recrimination and loss of face. The objective of criticism should be to prevent a recurrence of the problem and not to punish the guilty.

Engineers should improve their interpersonal skills to be more effective not just as professionals but as contributors to important technological decisions that increasingly drive social and economic development. It is sad that engineers, so uniquely qualified to guide us in this increasingly technological age, have not acquired the personality and skill required for effective decision making and leadership.

ENGINEERING ORGANIZATIONS

There are nearly 100 different engineering organizations, associations, or institutions in the United States and somewhat smaller numbers in other Western countries. Most of these are designed to organize narrowly specialized engineering professionals for the purpose of providing a forum for the discussion and review of technical issues in their area. There are some larger organizations in the traditional engineering disciplines such as electrical, mechanical, and civil engineering which usually have ten or more specialized subdivisions. Since World War II alone over fifty new specialized organizations have been formed to cater to the narrower interests of some engineers. These new, as well as some older, organizations overlap in their areas of interest. As an example, "engineering management" is covered by divisions in all the major engineering societies as well as other specialized societies such as the Operations Management Society and the Institute for Management Science.

As mentioned before, engineering organizations exist mainly to provide an area for discussion and social activities. In general, they do not engage in

1. setting engineering standards,
2. lobbying for legislation favorable to engineers,
3. protecting and enhancing the status of engineers,
4. protecting engineers from liability,
5. helping to improve engineers' salaries and conditions of employment and work, or
6. setting requirements for engineering qualification and licensing.

In other words, they are engaged in technical but not in professional issues.

ENGINEERS AS ENTREPRENEURS

Entrepreneurship and the drive to innovate are inherently associated with engineering, yet their formal training usually induces engineers to be risk averse. Risk aversiveness introduces an irrevocable conflict that engineers try to overcome all their lives. They like the excitement and challenge of developing new technology yet are held back by the inherent risk involved in new technological developments. As a result, when engineers become entrepreneurs and get carried away by their technological dreams and business aspirations, they usually withdraw or are forced to withdraw soon because they are unable and unwilling to sustain the risks and uncertainty involved.

Many of our high-technology firms were founded by engineers, but few of these founders remain at the helm of their firms for long. The reasons for their short tenure are many. Most were unable to consider broader business concerns and focused too much on narrow technological issues. Instead of adjusting their solutions to the needs and identified concerns or demands of the market, many otherwise very effective engineers, for example, consider the market as something that should respond to their "obviously" desirable solutions. While opportunities sometimes exist to mold the market, ultimately the market leads to at least short-term technological developments.

Engineering entrepreneurs often get too enamored of their idea and lose sight of the market and broader societal concerns. They then are disappointed when society and the market do not recognize their contribution, and they often resort to blaming others for their failure.

Engineers could be effective entrepreneurs and help resolve important problems if they were trained and able to

1. weigh risk effectively and learn to accept risk as inherent in all human activities, particularly in activities that deal with something new;

2. consider the broader implications, uses, and concerns associated with the introduction of changes in technology;

3. understand that society and markets accept only those changes that make social and economic sense in both the short and long run; and

4. make technology user friendly and adaptable.

Engineers could and should be the world's premier entrepreneurs who lead us out of our manifold technical, social, and economic problems, many of which could be solved with new technological approaches. But to achieve this goal, the perception of engineers must change and become broader, focused without losing sight of the general issues. Engineers must learn to accept feedback not as criticism but as productive suggestions designed to make their technology more acceptable.

ENGINEERS AS RISK TAKERS

Engineers are highly pedantic and usually attempt to assume absolute accuracy. They will hold on to a design or other problem as long as possible in order to refine their solutions over and over. In deriving their solution they use experimentally, empirically, or analytically derived approaches which usually encompass or make a number of assumptions. In many cases, engineers will converge on the basic problem solution rather quickly but then spend much more time converging on their final results. Even then they are quite uncertain of the reliability of the solution and will apply something called a safety factor—more appropriately called a risk aversiveness factor or a factor of ignorance—to adjust their result. The safety factor is usually a number such as 2 or 3 by which the end result is multiplied to take care of all the risks, uncertainties, and assumptions included in the problem solution.

The unfortunate thing is that most engineers do not know where safety factors come from, how they are derived, or what they are designed to correct. As a result, safety factors usually are taken out of the air and applied universally to all aspects of the design or solution, independent of the risk involved in a particular design result. It is curious that engineers who spend hundreds or thousands of hours refining an engineering solution will spend no time at all analyzing or justifying their choice of safety factor, even though the application of this factor will modify the results more than all the engineering work from the very initial results to the final outcome.

While it is nearly impossible to assure complete knowledge of all factors and assumptions that go into a solution approach, it should be possible for engineers to determine and justify the level of safety factor required. Statistical methods could be employed to estimate the different safety factors to achieve a consistent degree of safety in a design.

DILEMMA OF TODAY'S ENGINEERS

The explosive development of technology and, in particular, information and communications technology in recent years has caused radical changes in industry and human lifestyle. It has affected humankind in many ways, most importantly by the emergence of service industries as the principal growth sector in modern economies. These sociotechnical systems which provide health care, recreation, energy, communications, education, transportation, and so on dominate today's society in terms of employment and their impact on everyday life. As these systems use more and more advanced technologies, their role and the need for people in their operation changes. While technology serves humankind, and in a way contributes to the increasingly higher standard of living, it also causes major new dislocations of societal interests.

Engineering service technologies require a different breed of engineers. In the past, engineering was concerned primarily with design, and most engineers were trained and worked as design engineers. Engineers today are required to integrate design into manufacturing, into operations, and into services. In the future most engineers probably will work as technical integrators or integrative and operational engineers rather than work in design, which can be performed by or with computer aids. Engineers in the future increasingly will become involved and responsible for producibility, operability, maintainability, reliability, and performance of products and services and not just for their engineering design in the narrow engineering sense.

This new demand offers engineers a new role, which could lead to new status, and a greater ability to influence, guide, and make important decisions. But a new breed of engineers may be required.

The new engineer will have to be better educated in the broader issues affected by technology, without compromising his or her technological competence.

10

Toward Technological Excellence

In the United States scientists have maintained their inventiveness and continue the stream of scientific discoveries that have made the country the primary source of basic knowledge for the advancement of technology. U.S. scientists habitually win the majority of scientific prizes such as the Nobel Prize, and the United States is universally recognized as the world's scientific leader. Its laboratories have a world-class reputation, as have many of its universities.

The same used to be the case in engineering; even now some institutions such as the Massachusetts Institute of Technology (MIT) and the California Institute of Technology (CALTECH) remain major magnets for technology. But things have changed. Over 35 percent of graduate students in the United States are foreigners and an even higher percentage of advanced graduate degrees are earned by foreign nationals.

Most important, most of the engineering degrees awarded today are in engineering science. Few engineering degrees are awarded for creativity in engineering or creative engineering problem solving. Creativity somehow is moved on a back burner. There still are many creative developments in various engineering disciplines, but most, where they occur, are left in a state of scientific discovery, and the innovation process which creatively moves a discovery toward useful application is somehow ignored and not fostered. Engineering creativity, in other words, is not encouraged, and future engineers are not trained to be creative engineers but to be scientific or engineering discoverers, or engineering designers.

Creativity in engineering still exists, but it is mainly among self-trained, self-made engineers. Our educational institutions with rare exception follow the pass of scientific rigor and engineering discipline, which may result in scientific breakthrough but seldom results in creative engineering applications. It is essential today to assure continuity from scientific discovery to creative innovation, imaginative application and effective integration of engineering development, operability, usability, reliability, and performance of new technology. Engineering no longer can concentrate on inventing and then designing products or processes. Both depend on each other, and their development is an ever-converging spiral in which product and process engineering search continuously for improvements toward the ultimate product designed for effective producibility, quality, and performance. Such an approach can significantly reduce the time from discovery to product or process perfection and thereby beat the ever-present race against obsolescence. Obsolescence is a serious hazard amplified by the accelerating march of technology. Technological excellence today is no longer a strategic objective but a necessary goal for economic survival of any industrialized nation.

PLANNING FOR TECHNOLOGICAL EXCELLENCE

Planning is often associated with central control and is disdained by many Americans, particularly if it involves a central authority such as the federal government. As a result, we have few economic, social, or technological plans. The United States, for example, is one of only a few industrialized countries without an industrial policy or development plan. There is little or no planning of national transportation, health care, or even education. Somehow we associate planning with socialism or communism and the discredited performance of these systems. Planning also is supposed to infringe on personal and individual freedom. Consequently, whatever planning is performed by government is usually general in nature and designed to assure a maximum of flexibility to individuals and groups.

This approach to planning misses the point and purpose of planning. Planning is not designed to control but to help and to orient. It does not have to direct but to assist in developing a consensus and sense of purpose or objective with an associated set of suggestions on how the purpose or objective might be achieved.

Planning is essential if groups of people are to cooperate and coordinate their efforts toward an agreed-upon goal. We usually have little difficulty agreeing on goals, but to achieve them requires some planning.

Planning is associated with vision. To be effective, plans should be creative and keep the vision and the related goals in mind. While planning, certainly at the federal level, is somewhat disdained in the United States, it is not disdained in Japan. The Japanese government has since 1957 coordinated policies and plans among government departments by the use of the Economic Planning Agency. This agency drafted general plans for Japan's economic growth and guided various government ministries in their planning. Education, particularly engineering education, was selected as a foundation designed to permit Japan to compete economically. Technological sectors of the economy were targeted for growth and assistance, and plans were developed to assist them with education, research, and economic support.

The Japanese plan was not just to target certain industries and related technologies and provide them with a source of professionals and other assistance; it was to prepare and make them more readily adaptable to changing economic conditions and technological environments. The plan was fostered by the close coordination of government-industry policy and to some extent decision making, including encouragement of cooperation among companies or corporations in areas such as new product and process technology development. The government and the leading business organization (Kendandren) would encourage joint planning and cooperative research and development up to a stage where participating companies could take the R&D results and develop them into a new, often proprietary, product. Ultimately, all participating companies would compete on the basis of the additional innovations introduced, their prowess in marketing, and cost or price control. This approach allows the Japanese to plan, coordinate, and cooperate in most basic technology development, which in turn results in great cost and time savings as companies with government participation pool not only financial and human resources but also their technology base and market demand projections. This approach is partially justified by the need for Japan to develop increasingly advanced technology and thereby added value as the country relies mainly on its human resources for economic growth.

Although in the past Japan has relied largely on adopting or acquiring knowledge or technology from other countries, it now concentrates on developing its own advanced technology and on applying or transforming knowledge into new technological products more efficiently and more effectively than others. As the Ministry of International Trade and Industry's (MITI) policy advisory board, the Industrial Structures Council, notes, Japan in the future must be more creative and must focus on original high-technology development.

Planning continues for technological excellence by giving direction to technology development at the highest government and industry level while restructuring the educational system to respond to future needs. Planning for technological excellence in Japan is fairly focused by the selection of a limited number of areas of concentration. Other areas are maintained but are usually assigned the role of following technology.

An example is the shipbuilding industry which led Japanese industrial revitalization between 1950 and about 1970; as an assembly industry it provided a controlled base load market for other industries such as steelmaking, electrical manufacture, engine and equipment manufacturing, and more. As it was very labor intensive at that time, it also provided a large training ground for a skilled industrial workforce.

Today Japan is still the world's leading shipbuilder, but its shipbuilding industry is no longer a lead industry used to help the development of other industries; the industry has been able to maintain its competitiveness by rapid adoption of advanced automation and robotics technology, which allowed it to increase its labor productivity at a higher rate than the rate of increase in labor costs.

This success story did not happen by chance; Japanese shipbuilding, which has purchased and installed in excess of 7,000 production and assembly robots since 1983, was guided to a large extent by joint technological planning of government and industry. As a result, inefficient facilities were shut down and the surface capacity of the industry was reduced by nearly 40 percent in terms of facilities. Yet the improvements in productivity actually allowed the industry to maintain capacity and reduce costs, while becoming an important proving ground for Japanese high technology.

Planning for technological excellence is necessary to give direction to technology development and thereby eliminate wasteful efforts. It not only permits industry to orient its efforts, but it also facilitates effective development of the educational and other support developments necessary in assuring the success of new technology development.

JAPAN'S APPROACH TO TECHNOLOGY TRANSFER

Although Japan has been a major beneficiary of technology transfer imports and has for many years taken advantage of the openness of Western technology development, including access to research and educational institutions, Japan apparently is less open in providing access to its research facilities, laboratories, and educational institutions. Recent visitors from both U.S. and Korean research centers reported that most Japanese research labs do not permit guided tours or open visits of facilities and certainly not detailed discussions of ongoing research or inspections of actual research setups or experiments. Instead, visitors are led to a conference room and given a prefiltered overview which is usually so general as to be useless. In fact, the Japanese attempt to gather more information from the visitor than the visitor is given, by impressing on him or her that his or her treatment has been special. Most foreign experts complain that the Japanese are stingy in terms of technology and knowledge transfer, notwithstanding the fact that Japan got most of its technology from America and Europe.

Some feel that Japan uses technology as a commercial and economic weapon and will extort any price for its technology. Japan is resisting the transfer of its technology to developing countries, even in low-technology areas in which Japan is no longer commercially interested. An example is the building of simple ships or certain consumer appliances where Japan prefers to export kits for assembly rather than manufacturing technology.

JAPAN'S DEVELOPMENT

Japan faces the problem of an increasing labor shortage. In transportation, construction, and manufacturing the average age of workers is now over 44. As Japan manufactures more advanced products, it requires more skilled and trained workers. It is estimated that by the end of the twentieth century well over 4 million professional and skilled manufacturing jobs will remain unfilled—notwithstanding an accelerating trend of exporting manufacturing jobs abroad.

Japan remains acutely aware of its vulnerabilities, not only its inadequate labor supply but also its dependence on other nations for food, energy, trade, and, most important, defense. Japan lives by trade and has become the world's most efficient producer and trader.

Japan not only has become a confident nation of people proud of their achievements, but it finally has shed its old perceptions and inferiority complexes. Its new image is reflected in the assertions of Japanese leaders and by the self-confidence of its people, particularly of young Japanese.

Japanese who remember the criticism of their products as cheap copies of U.S. technology in the 1960s are now shaken by the decline, or at least relative decline, of U.S. products which provided standards for Japanese manufacturing for so long. They now assume that U.S. goods do not sell well because they are poorly designed and poorly made, not because of trade barriers. They also assume that U.S. management and marketing is basically incompetent. They similarly disdain the supposedly low level of discipline in U.S. industry, particularly among its workforce.

The United States will overcome this new perception not by negotiations, threats, or verbal defense, but by reestablishing U.S. technological excellence, not just in scientific but in real technological terms. To do this, we will have to learn to integrate all phases of technology development, reeducate engineers, restructure industrial management, give technologists greater responsibility, and eliminate the confrontational relationships among government, industry, labor, and society. We must learn to cooperate and coordinate our efforts, set priorities, and adjust our social and educational system to support these priorities. The path to technological excellence requires integration, collaboration, and cooperation, not confrontation.

INDUSTRIAL POLICY AND ENGINEERING REQUIREMENTS

One question that often arises is whether it is necessary to imitate Japanese industrial policy to achieve economic success or to hold on to our economic position.

Japanese society, lifestyle, organization, and, most important, government are distinctly different from those of the United States and the Western world at large. Even though Japan is a democracy in theory, it still is very much centrally planned and its economy is largely guided by government policy. Japan has a strong central government that backs a committed professional bureaucracy that in turn guides industrial and economic growth—including the development of increasingly higher technology.

Some authors such as Dietrich (1991) claim that to reassess our economic position we may have to change the way we organize and govern ourselves. Otherwise we well may become a second-rate nation, controlled by the Japanese.

Yet at this time, with Japanese industry gaining greater shares of the world market, there are signs that Japanese society objects to remaining docile and responsive to the edicts of senior bureaucrats. With Japanese industry's move offshore, there are new indications that to succeed in global competition as producers not only for, but in, foreign markets Japanese corporations must become more international, even Western. This pressure has resulted in democratization of major Japanese industrial concerns, under pressure by Japanese executives working extensively abroad.

As Japanese travel more extensively, they similarly demand more of the goods and quality of life that well-to-do Westerners take for granted. We can expect increasing pressure on the Japanese bureaucracy, from expatriate Japanese managers and an increasingly more demanding domestic workforce, for a more democratic Japanese society.

The docile, responsive, responsible, loyal, lifetime employee of major Japanese corporations probably will be replaced by a more educated, responsible, yet demanding employee, concerned as much with his or her own quality of life as with the good of the corporation and Japan as a nation.

The democratization of corporate Japan will not diminish the commitment of its employees or necessarily their loyalty. It will, though, cause a change in the traditional structure of Japanese corporations; more important, it will change the relationship between the government bureaucracy and corporate management. This process will be hastened further by recent revelations of corporate wrongdoing often involving senior corporate and government officials.

THE ROLE OF ENGINEERING EDUCATION IN AIMING AT TECHNOLOGICAL EXCELLENCE

As noted before, the percentage of engineers with graduate degrees in the United States is smaller than that of architects, chartered accountants, doctors, or lawyers. Less than half the engineers with a college education and less than 30 percent of professional engineers have gone to graduate school. Nearly 21 percent of professional engineers do not possess a degree at all.

Part of the reason may be that an undergraduate degree was for long considered a basic qualification for an engineer, and differences in starting salaries between incoming engineers with a bachelor's degree were only 10 percent below those with a master's degree ($35,000 and $38,200, respectively, for new mechanical engineers in 1991), which did not make the investment in time and money for graduate study attractive. Another reason may be that graduate engineering education is less and less aimed at improving engineering skills and is aimed more toward the training of research assistants, future researchers, and teachers. For a student of engineering who wants to become an engineer, graduate study may not be very attractive.

At the same time, universities encourage the brightest students to go on to graduate school, which in the U.S. environment often results in such engineers progressing toward a research or academic instead of an industrial career. The percentage of engineers with graduate degrees working for industry is comparatively small, and by some estimates it is less than 25 percent. This fact would not be so troubling if engineers with only an undergraduate education had ample opportunity to enhance their knowledge and education.

In Japan, most engineers hired with bachelor's degrees by major industrial firms (who hire about 66 percent of all graduating engineers) are given an average of 1 year of additional training during their first 4 to 8 years with the company. The brightest may be sent abroad to the United States or Europe for graduate study, while others usually will receive additional training at in-house training institutes, some of which are similar to small research and teaching universities. (Continued education is offered throughout the career of a professional engineer.)

Comparatively few engineers working for industry in the United States and Western Europe return for graduate study or are given opportunities for substantial training during employment. True, many are sent to short courses of a few days' or weeks' duration, but these short courses usually are oriented toward honing a specific narrow skill of the employee and do not provide further engineering education. Furthermore, the sponsorship of engineers for such courses usually takes place in connection with a specific job requirement. In the United States many engineers are not expected to spend a significant part of their professional life with just one employer; consequently, employers do not consider it attractive to invest in an engineer's education, without any plan or commitment from either the employer or employee that the relationship will continue indefinitely.

In some Asian countries such as Singapore, and to a lesser extent in South America and Europe, a system of bonding is used. Under such a system an employee undertakes to serve a given number of years for an employer for each year of study (usually graduate study) funded by the employer. Under such conditions the employer not only provides tuition, travel, and upkeep during the term of the study but also maintains the employee's salary and benefits payments.

Employers must be convinced that graduate study enhances the professional competence and contribution of an engineer before they will be able to fund graduate study or pay a meaningful differential for a graduate engineer. This will happen only if and when graduate engineering study is directed toward the needs of industry and not academic research. A graduate engineering education must concentrate on relevant engineering subjects and engineering problems that are of social and economic significance.

Unless our graduate engineering schools move in this direction, they will become even more isolated ivory towers that may find it more and more difficult to obtain the support of government, society, and industry. In recent years the relevance of U.S. and, in some cases, Western university education and research has been scrutinized and in many cases questioned. This scrutiny will continue and mount unless universities themselves reevaluate what they are doing.

There are many who now believe that universities often are self-serving in the way they run and finance education and research. It is important for universities to reevaluate themselves critically and to make the necessary changes to prevent a radical change in their role as perceived by the public and ultimately the government.

Engineering education, particularly graduate education, should be focused on the needs of an excellent professional engineer, a true professional, who can solve problems of technology, advance technology, and make technology serve the needs of humankind. The focus should not be on the needs of a professional who is given problems to solve and looks at problems in a narrow sense often defined by nontechnical people.

Engineers must learn not just to be creative in a purely scientific sense but to enhance their creativity by the rapid innovation of technology into useful products or processes. The Japanese have trained their engineers and have developed organizations that assure significantly faster technology innovation and development than occurs in the United States. As a result, the Japanese not only advance technology more rapidly but also bring it to market and win market share more effectively.

Earlier in this century the United States enhanced its technological world leadership by adopting and developing inventions from Germany and Great Britain. U.S. engineers turned creative foreign as well as U.S. scientific inventions into advanced products and processes more rapidly than others could. The American engineering genius at that time developed new products and processes for use and rapidly produced them for a market bred on technological innovation. American industry was for decades supreme in its ability to transform basic inventions into useful readily marketable goods, processes, and even services. Somehow this ingenuity was lost when large-scale mass production took over and cost, and consequently price, became the market generator instead of technology.

But the table has turned again. Technology moves now so rapidly that nobody can afford the long-term development and massive investment in mass production. Instead we are moving toward shorter-term technologies produced in smaller batches in flexible manufacturing enterprises readily capable of changing products and directions.

While U.S. manufacturers have started to change their plants and methods of production and many have introduced large-scale automation or robotics to permit rapid changeover from product to product, engineers and engineering departments have been less responsive to changing conditions. Products now must be developed more quickly and perfected to be producible at high quality in a very short time. This process requires not only creativity in design and engineering but integration of production and user requirements to assure that the new product not only can be built as a prototype but is producible in desired quantities with high quality right from the start.

The engineering skills required for this process go beyond any currently taught to U.S. engineering students. In some countries, such as Germany, apprenticeships fulfill part of this role and are encouraged. In Japan, engineers spend part of each year or two in manufacturing, marketing, planning, or research. Something similar must be provided in the United States if our engineers are to acquire the desired combination of scientific knowledge, engineering capability, creativity, integrating skills, and an understanding of not just how to design and engineer products and processes but to develop, integrate, produce, and make products and processes work for the user.

Engineering education needs the help of practicing, experienced engineers, particularly engineers with broad design, manufacturing/operating, and marketing experience. Few engineering faculty have industrial experience, and by some estimates less than 20 percent of U.S. engineering faculty ever worked outside a university except as consultants or on short-term assignments. Inbred engineering faculty in the United States lack a perception of the needs of the outside world and the expectations by industry of what constitutes a good engineer. As a result, engineering education has become self-centered and largely replicates the self-perception of the faculty, who know little if anything about what engineers do and how they work. Engineering education, particularly at the graduate level, must include real-world problem solving, and teach students how to innovate and how to assure producibility of their designs.

The introduction of courses that educate engineers in problem solving and innovation also can serve engineers who choose to become engineering researchers and educators by assuring that they know how to make their work more relevant.

Even if a practical apprenticeship is unrealistic as part of today's engineering education, at least internships of 3 summer months every year during graduate and possibly the senior year of undergraduate study is not unrealistic. Such internships not only would provide future engineers with an understanding of the job environment and the reality of engineering, but they also would permit the engineers to focus on their own particular interests. They would provide engineers with researchable ideas and networks of contacts. Overall, they would make engineers well-rounded, confident, and balanced people capable of performing as true professionals.

Many may consider an internship, providing industrial and particularly shop floor or design office experience, demeaning for engineers. Others may consider it unnecessary and outdated. Yet all probably will agree that if graduating engineers are to play a greater role in the effective and timely development of world-class technology that will stand up to any competition, future engineers will have to move from their computer workstations, computer graphics screens, and laboratories and smell the odors of the production floors, hear the comments of technology users, and understand the issues faced by planners and managers. Engineers will have to learn what makes technology work, how it can be put together, and how it can be made to work.

Engineering education no longer can end with graduation and rely on a few short skill courses to maintain an engineer's skills. Engineering more than other professions requires continual updating; technological change is at least as, if not more, radical and dynamic as change affecting other professions.

An urgent need exists for advanced engineering education where engineers return to school every 5 to 10 years for a few months a term or even an academic year to upgrade their knowledge. This approach would benefit not only engineers working for industry but also faculty in schools of engineering and their graduate students who would be able to interact with practicing engineers, who would bring real-world problems and issues into the classroom.

I would propose that professors of engineering be encouraged to spend their sabbatical leaves not at research institutions or other universities, where they will perform basically the same research and other activities (though at a slower pace) as at their home institution, but at an industrial plant or with an industrial firm engaged in an area in which they educate engineers. For engineering education to become and remain more relevant, engineering faculty members must be exposed to the realities of engineering, the industry they serve, and the technology and its use they claim to contribute to. Unless such steps are undertaken, engineering education will become increasingly academic.

Relevance and total quality management (TQM) of engineering education must go hand in hand. In engineering education TQM implies the efficient, just-in-time introduction of subject material to serve the program objectives. Engineering curricula should be structured to build engineering skills logically on a foundation of basics. Engineers must learn not only basic science and engineering but also contextual, management, interpersonal, and project management skills. Educational quality management assures that courses or subjects are effectively sequenced without gaps or overlap between sequential subjects. Subjects should complement and build upon each other not only in narrow focused concepts but also in contextual terms. They should train engineers both in

scientific principles and engineering theory and in problem identification and solution, in which principles and theory become one
of many essential tools. Environmental and social issues also
should be addressed by engineering curricula.

Engineers must learn that engineering is part of the technology development process. Just-in-time engineering and the
management of engineering activities are important elements
of modern management of engineering education. Engineers
must learn that, while perfection is always a goal, it may not be
achievable without building in obsolescence into otherwise
good engineering solutions. In other words, engineers must be
taught that a good and timely solution is often more important
than a perfect solution offered too late to make a difference.
Teaching quality in engineering includes the understanding of
an effective balance between good engineering, engineering
economy, and engineering timeliness.

Engineering management, for example, should cover not only
the economic and financial cost and benefit of engineering but the
management of engineering itself. The effectiveness of an engineering solution not only is measured in terms of the economic
success of the solution but includes the economic effectiveness of
the engineering process itself. Engineers must learn to ask if the

- engineering effort and time is justified by the problem
  addressed,
- value of the solution changes over time and how it is
  affected by time,
- trade-off of time and engineering effort has been effectively
  made, and
- engineering activities are planned properly to assure just-in-
  time delivery of intermediate and final engineering solutions.

In other words, engineers must learn to manage engineering.
Engineering is not an experiment but an integral part of technology
development in product and process projects where the effective

balance in the performance of all required activities is essential for success.

Engineering management courses should include instruction in managerial behavior and leadership. They must cover the use of authority, the ability to influence, project organization and management, project planning and control, as well as the effective evaluation of risk in engineering and risk-weighed decision making. Engineers must learn to confront uncertainty in a rational, analytical manner and must not hide from it by the use of unsubstantiated safety factors.

A major reason for the narrowly focused approach engineers seem to take in the tackling of problems appears to be our emphasis of analysis in engineering education with the nearly total exclusion of synthesis. Engineers are trained to be investigators and problem solvers in the narrow sense of specific engineering problems. Engineering education in the future should emphasize both analysis and synthesis, thereby assuring that engineers understand problems and not only acquire the skills to solve problems defined or identified by others.

ENGINEERING EXCELLENCE

Excellence in human endeavor is a requirement for quality in life. To attain excellence, a person must care more than is considered wise; risk more than is considered safe; expect more than is considered possible or reasonable; and be willing to think about things considered impractical, unusual, or nonconforming. These characteristics used to define many engineers in the past, but no more. Engineers are perfectionists in many ways, but they usually are too focused to be inventive; too risk averse to challenge known, practical approaches; and too narrow to be truly creative. Engineers are very good at making little advances but often fear taking bigger steps. They will do things with which they are familiar but rarely will venture into unknown territory. This is not a recipe for excellence. At a time when humankind faces new problems, many of which are the result or effect of technological change, there is an urgent need for engineering excellence in the broader sense as defined above and not as the narrowly focused definition of scientific quality. Engineers should be competent, effective, and true users of science, but their calling is to use their ability to make science work for the good of humankind and the well-being of the earth's environment. Science for science's sake is for scientists. Engineers should be doers with a vision, creative instincts, the ability to recognize needs, and the ability to match or develop technology to meet these needs.

WHAT IS AN EXCELLENT ENGINEER?

How does one measure the quality of an engineer? This question is difficult to answer as engineers perform many different functions. An excellent engineer today should

1. know engineering principles and know how to apply them effectively;
2. be creative and entrepreneurial;
3. have a bias for action;
4. know how to communicate;
5. know how to be productive and assure the productivity of others;
6. understand users of the technology or services he or she is working on;
7. be hands-on value-driven;
8. know how to deal with people;
9. understand his or her limitations;
10. be willing to judge and assume risks;
11. be decisive and reliable;
12. be supportive of others, particularly other engineers;

13. stand behind his or her professional work and judgments;

14. be willing to learn;

15. be conscious always of the need for quality;

16. understand how to integrate user needs, operability, maintainability, and producibility into the development or design of a technology; and

17. understand the economic, environmental, and social implications of engineering solutions and know how to assure a most effective technology.

These characteristics make an excellent engineer a well-rounded, applied scientist who is able to solve problems and respond to human needs and opportunities effectively.

ENGINEERING FOR THE PUBLIC AND THE WISE USE OF ENGINEERING

Alfred Keil, the former dean of engineering at MIT, was an early proponent of engineering for the public and the public responsibility of engineers. He formulated the phrase "wise use of engineering" and developed the concept into recommendations for improvements in the teaching and practice of engineering. Keil advocated that social and environmental objectives should form an integral part of engineering project and design objectives.

We face difficult resource constraints and environmental problems for humankind on this earth. While some problems are the result of an increasing population, which may reach 10 billion before the end of 2055 and 20 billion before the end of the twenty-first century, many of the problems we face today, both in terms of resource availability and the condition of our environment, are the result of the impact of the unwise development and use of technology. As an example, carbon oxide gas emission into the atmosphere from 1965 to 1990 equaled the total emissions during the preceding 5,000 to 8,000 years of human civilization.

Similarly, over 50 percent of fluoride gas emissions, which reduce the protective ozone layer, have occurred during the same period in time. Similar problems exist on land with groundwater aquifers, rivers, coastal zones, oceans, and the earth's crust itself. Some pollution of space has started as well.

In many ways we often introduce technology for technology's sake or for narrow profit, strategic, or competitive reasons, without a clear indication of actual need. The need for technology today quite often is derived. While technology probably enhances the quality of life on balance, there is much bad technology around, which contributes little to human well-being, while consuming valuable resources and often harming the environment.

Technology quite often is a fad that comes and goes, contributing little if anything of value. Engineers by and large have not been involved or interested in issues of technological relevance and impact. They left the moral, ethical, resource use, economic, social, and environmental questions raised by technology to others. As a result, technology was and often is misused by those who claim that they do not understand technology. Engineers cannot leave these issues totally to others. They cannot wash their hands completely from the misuse or detrimental impact of technology. They are responsible for the technology they develop and as such for how it is used.

Technology affects society and often fosters social change. Many examples exist where technology or technological change upset sensitive social balances and where technology was misused. Engineering serves society by developing uses of scientific knowledge that improve life and living standards. Unfortunately, acceptance of technology often is mired by lack of understanding of social, economic, cultural, and political barriers. There is an urgent need to educate users of technology while assuring that people will not misuse partial knowledge as do-it-yourself technologists. Ultimately, engineers are blamed even when they share no responsibility for the use or misuse of the technology. Their association with the development of the technology usually is enough to justify assignment of blame.

This issue is becoming more serious now when technology affects larger and larger areas or environments. Technology increasingly affects not only direct users of its immediate environment but more often the larger community. Similarly, technology often is used for other than the intended uses, often with catastrophic results.

Engineers no longer can wash their hands and claim that their involvement is limited to the development of technology. Technology has become too complex and, in many cases, potentially hazardous to let decisions on how, where, and for what to use it to be made solely by nontechnical people. Engineers must assume a role in decisions affecting technology applications and use. They must define the skill required to operate the technology and the purposes for which it should be applied. For this, engineers will have to attain a better understanding of the market with its technology users, sellers, and the environmental and social conditions under which the technology might be employed.

Engineers will have to lead the wise use of technology by designing technology that can be effectively used only for its intended purposes. They also will have to take an active role in guiding or leading technology development and use policy on a national and global level. Here again engineers have, by and large, been timid. Few have assumed decision making positions in national or international policy forums. As with national governments, international organizations that guide environmental, energy, and similar policy-making bodies usually are headed not by engineers or other specialists but by politicians, lawyers, or bureaucrats. Only two out of ten United Nations bodies concerned with technical policy making and enforcement are headed by engineering specialists. The reason appears to be that few engineering experts have or are willing to exert the leadership necessary to head such agencies and assure the formulation of wise policies.

The problem revolves around the conflicts among technological usefulness, profitability, environmental impact, social and political advantage, and strategic or status benefits/costs. Engineers usually are neither good decision makers nor effective negotiators. Even though they may be able to judge the costs and effects of a new technology better than others, they are at a loss when leadership and decisiveness are required. Even though the Superpower confrontation has evaporated with the demise of the Soviet Union, the proliferation of weapons of mass destruction and environmentally hazardous technologies abound. Engineers will have to assume a leading role in returning this world to sanity and the wise use of engineering, as they are the ones who put the devils in Pandora's box in the first place.

11

Model of the Engineer
of the Future

The role of engineers, at least in the United States, will have to change radically. Engineers will be not only the technology developers, but the guides who lead society through the maze of technological developments toward a brighter, safer, environmentally cleaner, and better future.

Yet today's engineers are neither trained for this function nor ready to perform it and assume its responsibility. If we think seriously about a "New World Order" in geopolitical and economic terms and consider its technological, social, and environmental impacts, then a new model—the engineer of the future—emerges: an engineer who takes responsibility and considers not only the technical and scientific but also the economic, social, environmental, and moral implications of technology. Only by considering the broader impacts of technology and by learning how to develop technology in broader terms will engineers become true leaders. Engineers today are largely followers who create, design, develop, and produce technology to satisfy the demands and needs of others. They try to be popular and fall prey to a desire to please. They associate success with popularity with their superiors and the users of their technological developments. They seldom follow up on their technological developments, and they do not insist that users be effectively trained or that technology be effectively maintained.

Yet engineering responsibility does not end with turning over some research result to design or an engineering design to manufacturing or even a manufactured product to the user. It continues for the life of the product or the technology. Leaders in total quality management have preached this philosophy for some time and have made progress in some quarters. But their success has come largely in assuming integrated quality management in engineering design and manufacturing.

Engineers are judged to be excellent by virtue of the quality of their choices, the effectiveness of the technology they develop, and the creativity of their decisions. Engineers are what they develop and are formed by their work.

Excellence in engineering is a judgment not on the sophistication or the elegance of the methods applied but on the contribution an engineer's work makes to the quality of life and well-being of humankind and the earth's environment. Excellence relates to quality and creativity and ultimately measures the relevance and impact of the contribution made. It is also a judgment of the life and extent of the impact made. Technological facts, although often impressive, are not signs of excellence in engineering.

ENGINEERING AS PART OF A TEAM

Teamwork by engineers is essential. Too much time and creative effort is otherwise spent on assuring consistency in material choice, dimension, production, and manufacturing procedures. Engineers usually prefer to work by themselves or in very small, specialized groups. Similarly, they are reluctant to share the evolving results and developing information with anyone outside their small circle. This reluctance is largely due to the fact that engineers feel that work is incomplete and will require changes. They often do not recognize that even information on their approach and the evolution of design is valuable to others and would contribute to better, more effective integration of all the engineering and research inputs, the product or process planning efforts, and the development of producibility.

The traditional approaches of configuration or design management used by many engineering organizations are no longer good enough because they consider each engineering change by itself. For example, if a dimension is changed in some part of a design, this change will be entered into the engineering design. Those affected by the change will have to discover it for themselves as the effects of the change on their work may not be immediately evident. In complex technology design this process may be ineffective as

1. changes made may not be discovered by those affected,

2. those potentially or actually affected may not recognize the implication of the change, and

3. changes may not be coordinated with product or process function, which is often aimed at a narrow purpose or use, affected by the change.

Research, engineering, design, production planning, manufacturing, and marketing must all be integrated and coordinated to assure a common understanding of the objectives and the expected or planned function, use, and application of the technology under development. For this coordination to be effective, engineers must mingle and exchange ideas and information with people from all groups associated or potentially involved with the technology under development. They should try out ideas for changes before making them. Conversely, they should solicit ideas and listen to the ideas of others often completely outside their narrow professional interests. They should try to envision all possible effects of changes made by others.

Teamwork in technology is not difficult. But people, particularly engineers, must be trained for it and must be shown that the efforts involved offer tremendous dividends and assure greater success for the technology development in time, quality, producibility, and usability.

The need for teamwork and the effective integration of engineering into all other activities is driven by a shorter product market life. The pressure exists not only to develop more products and processes but to develop them faster and with more reliability and quality. We no longer have the luxury of long lead times, market test periods, and market feedback. As shown in Figure 10, we now need larger integrated engineering teams that work closely with marketing, research, product development, and manufacturing.

Shorter product life cycles and the more rapid emergence of new technologies pose the problem of technological catch-up where, for example, a new product technology, because its development was too slow, is overtaken by a newer technology (see Figure 11).

In Figure 11, technology A takes 6 months to be invented and another 30 months to be developed to about 95 percent of achievable perfection for a total of 36 months, after which it reaches maturity and subsequent obsolescence. Technology B follows by 4 months and also requires 6 months to be conceived, invented, or discovered. But its development is much more rapid, and it reaches 95 percent of perfection after 28 months, or 18 instead of 30 months from discovery. In fact, technology B overtakes technology A in month 16, or 6 months after the start of its development phase. Under these circumstances, which occur more and more frequently today, the overtaking technology will nearly always capture increasing market share from the preceding technology whose development rate will, as a result, decline further because of talk of market support. This is a dilemma for many engineers who like to hold on to engineering problems and rework them indefinitely.

Quality in engineering no longer is simply concerned with the quality of the analysis made; in the future it also will involve the timeliness of the solution. A more perfect solution that is later and can no longer be used is not a sign of quality or excellence in engineering.

Figure 10
Engineering Trends

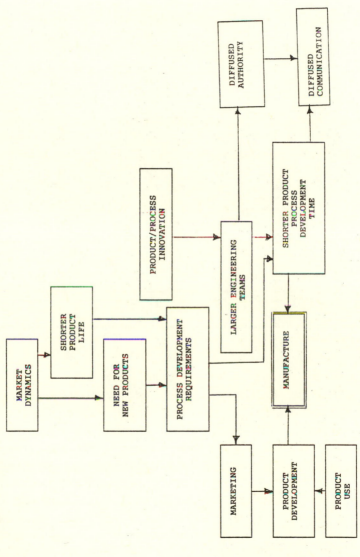

Figure 11
Overtaking Technology

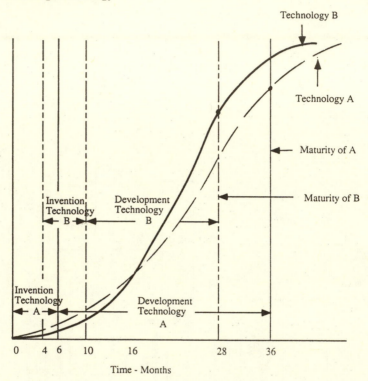

Time - Months

As the depository of technological knowledge and guardians of technology developments, engineers must assume a greater leadership role, both at the team and at the public level. They must become more influential not only in the selection of new products and processes but also in setting company and public or government technology development policy. Many claim that engineers are too rational to set policy and that effective policy making requires dreamers and people who can look beyond the realm of rational consequence and logical development. Engineers must learn to relax their rational thinking and formal step-by-step approach and free themselves to imagine futures not solidly built on the past. Engineers must correct many of the mistakes of the past, introduce unorthodox solutions, and consider new approaches to technological development. Quality in engineering then will be measured not just by the degree or sophistication of analysis and design but by originality of thinking, creativity of approach, and effectiveness of solution.

In many cases, simplicity will provide the superior quality in engineering. Quite often engineering complexity has overtaken technological rationality. More complex technology is not necessarily better, and complexity too often is driven by short-term marketing considerations. Similarly, complexity often discredits basically sound technology. Engineers should learn to resist unreasonable demands for incorporation of features that make a technology less reliable and otherwise downgrade or discredit its performance.

PRODUCTIVITY, QUALITY MANAGEMENT, AND ENGINEERING LEADERSHIP

Productivity and quality management are considered issues for the shop floor or production workers and the management of manufacturing. Traditionally, engineers did not recognize any role for engineers in productivity and quality management. Similarly, labor for a long time refused to consider it one of its concerns, unless a direct monetary incentive was involved. Slowly and usually with some suspicion and trepidation, labor, including organized labor, is beginning to accept the fact that workers as well as employers have an interest in raising productivity and quality. On the labor side, this has resulted in the elimination of many restrictive work practices, those that affect the loss of time as well as those that influence output and quality during the remaining time when actual work is being performed.

As a result, labor productivity has increased appreciably on many a U.S. shop floor. Labor is beginning to recognize that increasing quality increases productivity and that both result in improvements in working conditions. U.S. industry finally is getting the message across to labor that unless our performance improves, our standard of living may be in jeopardy.

Many approaches, largely imported from Japan, were used by U.S. industry to improve productivity. Among these approaches were quality circles, teamwork methods, leadership development, just-in-time delivery, and a full commitment to customer satisfaction. The problems often experienced in introducing these methods to U.S. industry were

1. lack of informality in interpersonal relations and in the discussion among quality circles,

2. hierarchical approaches to discussions with the senior staff,

3. involvement of only direct production workers and production management in most TQM efforts,

4. little, if any, engineering and marketing input,

5. insufficient management support, and

6. ineffective and often delayed implementation of recommendations.

The most important problem with making TQM work in the United States is the lack of involvement of engineering in quality and productivity improvements in many U.S. corporations. To some extent, this problem is due to the remoteness of engineering and management staff from the production floor. Equally important is the erroneous assumption that engineering plays little role in quality and productivity improvements. This last issue is the key to improvements in competitiveness.

The narrow focused skill of U.S. engineers served well when U.S. industry relied on mass production to manufacture goods inexpensively and hired or fired engineers as changes in products and processes were required. Engineers were specialists with narrow skills and generally could not respond to changing technological needs. Engineers were considered a cost center to be controlled when changes occurred. Consistent with this concept, engineers were organized in narrow specialized fields, each responsible for a specific area of design or development, without effective integration. This approach resulted in a highly structured engineering organization without flexibility or overview and too large an inertia to respond or adjust to changes from within or from without.

Engineers generally were given narrow responsibilities and few decision powers. Most important, their work was not integrated with manufacturing. With a narrow focus, engineers lost innovative skills. This approach caused much of the U.S. industry to become a low cost producer of standard goods produced in large quantities, requiring few changes. Changes were difficult to make and took a long time. New models of automobiles, for example, took 6 to 8 years to develop and bring to market in 1980, a period now reduced to 4 to 6 years.

More rapidly advancing technology made this approach increasingly noncompetitive, as U.S. manufacturers found it difficult to respond to changes in technology and market demand in a timely and efficient manner. The result is the well-documented loss of competitiveness and market share. Technology and quality have become the driving forces of a more informed and increasingly international market that, only a few years earlier, was driven largely by costs.

PEOPLE MANAGEMENT AND TECHNOLOGICAL EXCELLENCE

To achieve technological excellence, people have to become personally responsible for their work and for how and when they do it. Excellence is achieved through independence which in turn thrives on trust. When people are trusted in what they do and how they do it, including the management of their own time, they usually will respond with imagination and interest, resulting in improved creativity.

U.S. industry must address its labor-management relationships and reevaluate management methods. Management methods still are excessively bureaucratic and include too many reviews, meetings, and controls. The initiative of people is impeded and work becomes less interesting. Innovation, which depends on the climate provided, also is affected. Most people, and particularly engineers, are innovative by nature and have the training to determine the basic viability of the innovation. To assure innovation that prospers, the work environment must encourage diversity, foster networking and communications, and permit independence and freedom. As mentioned before, true innovation can be achieved only under conditions of trust and freedom where people take responsibility for their work, working hours, and work content. All management should do is set objectives, identify general needs to be met in terms of new products or processes, and establish general standards. In such an environment, people can be creative. They may not always succeed, but they will be motivated to do their best.

The goal of effective management is to unleash the potential of all the people. People usually are able to perform to their full capacity only in the right environment. If this environment is not fostered and maintained by management, creativity and performance will be compromised and the opportunities of achieving technological excellence may be lost. Today, when we strive to retain or regain world class technological standards, effective nourishment of our most valuable resource, people, must be given greater emphasis.

12

Technological Excellence and Engineering Leadership

Technology has many dimensions. It is not only a physical object capable of performing some function; it is also the methods or procedures used to accomplish some function with or without hardware technology. Knowledge and software, therefore, are as much technology as a complex piece of hardware.

Technological excellence involves creativity and innovation, quality in design and manufacture, operability and reliability in use, environmental compatibility, economic contribution, adaptability, quality in material (if applicable), maintainability, and more. Technological excellence implies achievement of the highest possible levels in all of these factors. It is not enough to develop a creative or innovative concept or technology; high performance must be achieved as well.

People have different opinions about what constitutes technological excellence. For some, it means a technology that performs beyond what is expected of it. Others demand that to be technologically excellent, a technology not only must perform well but must be beautiful, aesthetic, or elegant, as well as reliable and advanced.

Creativity is an essential requirement for technological excellence and engineering leadership. A National Science Foundation (NSF) report (1976) noted that during the 20-year period 1953–1973, the United States developed 63 percent of all significant technological innovations. This achievement was credited largely to American emphasis on origination versus the stress on pragmatism emphasized in Japan, for example. While the United States is still a leader in scientific discovery, it has lost much of its leadership in technology development due to the fragmentation of research, development, and implementation, as well as the lack of emphasis on the integrated quality management of the stages from discovery to application and marketing as a continuous process.

Individualism served U.S. technology development well when technology was simple and could be developed by individuals and when each technological innovation was unique and inventors or innovators did not have to worry about overtaking technology. Conditions today require effective teamwork and rapid technology development in a timely manner which assures that the technology it is not overtaken and that it works effectively and reliably when brought to market. This approach requires integrated teamwork and top-level support and cooperation among researchers, engineers, production experts, marketing people, prospective users, and industrial concerns with interest in the same technology.

Individualism on a personal and corporate level may not be appropriate when it comes to technology development, which now requires cooperation and collaboration on the individual and corporate level to achieve world-class technology development and technological excellence that can compete anywhere. The requirements for successful leadership by engineers in the development of technological excellence (without a definite ranking order) can be defined as follows:

- Creativity
- Vision
- Persistence

- Capacity and willingness to communicate
- Consistency and focus
- Ability to create cooperative environments
- Enjoyment of their role and their work
- Ability to delegate effectively
- Good at objective and goal setting as well as defining fair performance standards
- Fairness with toughness
- Good organizers
- Ability to influence and motivate
- Good at conceptualizing whole impacts and uses
- Commitment to make things happen

Engineers in the pursuit of technological excellence must question technological aspects but not express cynicism, mistrust, or suspicion of the work of an organization that employs them or individuals working with them. It may be equally important to refrain from cynicism of work performed by other organizations. Comments should be as direct as possible and consist of straightforward, honest evaluations. Technological excellence and engineering leadership feed upon each other and are mutually interdependent.

The United States has excelled in many areas. It excelled in exploration, inventiveness, courage, individualism, craftsmanship, and use of liberty or freedom in expanding human horizons during its formative years. Since reaching the pinnacle of world leadership in the twentieth century, the United States has concentrated its excellence in more focused areas such as scientific research, engineering innovation, medicine, human rights, politics, and art. This concentration may be a sign of maturity and self-confidence, but we now have become a nation of increasingly narrow specialists who are concerned with the outcome of their focused activity.

This narrow specialization applies to medicine where we have some of the world's greatest specialists who routinely win the bulk of awards like Nobel Prizes yet are unable to provide adequate health care to the average citizen. It applies to law where we make tremendous contributions to jurisprudence and have the world's largest, most complex legal system employing 60 percent of the world's legal professionals yet are unable to control or even effectively adjudicate crime. It applies to science and engineering where we lead the world in basic research and innovation yet are unable to transform these advances into effective new useful technology in a timely and effective manner.

We know how to develop technological excellence. We are leaders in the science of quality control, but we somehow lose the ability to apply it to the everyday needs of manufacturing, operations, and services.

THE LINK BETWEEN TECHNOLOGICAL EXCELLENCE, ENGINEERING LEADERSHIP, AND ECONOMIC GROWTH

Without engineering leadership we can achieve neither technological excellence nor significant economic growth. The period between the Civil War and World War I, a short 50-year timespan during which the United States established itself as the world's preeminent industrial and technological leader, was a time of unparalleled U.S. engineering prowess. It was a period of engineering ingenuity that had a profound effect on the standing of the United States as the unchallenged industrial and technological leader of the world. The United States moved ahead of the other leading industrial nations in practically all aspects of technology, from coal and steel production to electric power generation and manufacturing.

During that time the U.S. economy grew as never before and became the economic locomotive of the world. U.S. technological push generated worldwide economic pull. This was largely accomplished by a group of independent inventors and technology developers who, unconstrained by government or industrial organizations, used their genius to push technology and thereby the U.S. economy ahead.

Although the United States continued its technological leadership between the world wars and became the economic locomotive that pulled the economies of friends and foes out of the morass left after World War II, it has since lost much of its technological leadership and, as a result, its economic influence. Its easy spending consumers who provided the market for its own and, since World War II, for the world's goods can no longer afford to pull the world's economies. In the process, while the United States became the preeminent world importer, U.S. technological development started to decline because many of the imports relied heavily on technology exports. It may be that because of overconfidence in the U.S. long-term technological lead, few in the United States recognized the warning signs of increased foreign technological prowess and competition.

Similarly, U.S. consumers became increasingly fascinated with things foreign. The increased transfer of U.S. technology to foreign low labor cost manufacturers producing goods for U.S. companies for the U.S. market was probably the most important factor in reducing U.S. technological developments. It resulted in the large-scale transfer of not only technological know-how, but also of U.S. marketing techniques, including knowledge of the U.S. market base. In other words, we first established modern industrial facilities in other countries, then transferred U.S. process and product technology to manufacture goods there for the U.S. consumer, initially under U.S. labels, only to find that after all the knowledge was transferred, foreign manufacturers simply used it to invade the U.S. market. In the process, foreign competitors also learned the importance of product quality and technology excellence in winning market share, knowledge again acquired from U.S. experts and experience.

Similarly, foreign competitors learned that to gain markets and economic leadership requires well-developed technological leadership and engineering excellence, an approach the United States itself had used before to become the world's technological and economic leader. During the long period of its leadership though, many aspects of these issues had been forgotten or were submerged beneath the narrow interests of short-term gain and financial greed.

The United States has forgotten that technological excellence is the lifeblood of economic growth and competitiveness and that both depend on engineering leadership. The United States has let engineering leadership lapse, has underinvested in plants and in human resource building, has allowed quality and excellence slip for assumed short-term cost savings, has institutionalized technology development, and has made it subject to bureaucratic procedures.

Technological excellence requires strong research support, freedom by researchers to act on their gut feelings, and giving engineers a chance to innovate without administrative guidance and legal or regulatory control or interference. A climate that encourages creativity and provides incentives to advance technology and apply it, even if it leads to failure, is essential for technical excellence.

Engineers only thrive in a nonbureaucratic environment in which interpersonal trust reigns supreme. They must feel that their work and ideas are appreciated and cooperation is encouraged. Similarly, engineers must feel that they are encouraged to lead, to develop new technology, and to advance knowledge, and product and process developments. Only leaders produce excellence, and engineers need to feel that they are expected to lead in the development of technological excellence. For a long time, engineers have been told that their role is to follow short-term policies set by financial and marketing leaders who were often constrained by legal requirements.

Government and industry laboratories have become highly institutionalized and have constrained their technology development through their own direction and inertia. They neither recognize the moment nor act in a timely manner. Most important, they set the technology problems to be investigated and do not allow radical innovative ideas to flourish. Most U.S. technology development has been institutionalized even in academic environments. While the United States retained world technological leadership in the 20 years following World War II, it has since started to lose ground in many technological areas. The reasons for this loss are manifold and include the following:

- Overemphasis on military R&D with the bulk of R&D funding and the majority of top researchers employed in noncommercial or even commercial fields
- Inadequate and inefficient transfer of technology from government and military R&D to commercial organizations
- Lack of continuity in the R&D to product/process development; inadequate emphasis on product/process innovation or application development; great loss of time between invention and product/process innovation or development
- Short time horizon in technology development planning
- Cutoff or delay of technology development to protect existing, often inferior, technology
- Lack of cooperation in both R&D and technology innovation
- Ineffective, often outdated, technology development strategies, with narrow focus and short-term profit horizons
- Adversarial noncooperative attitudes and purposes between government and industry
- Overregulation, legal impediments, and excessive legal involvement in all stages of technology development and use
- Inadequate knowledge transfer
- Lack of effective training of workforce and narrow education of engineers
- Lack of knowledge of real problems and the social, economic, and environmental implications of technology
- Lack of encouragement and training for teamwork and intraorganizational cooperation

While the United States still maintains a high level of technology, it has lost much of its technology development momentum. This situation is largely the result of the fact that the United States is now as large a high-technology importer as exporter. Although it still exports as much high technology as Japan (about $94 billion in 1989), it imports a nearly equal amount, while Japan imports very much less. Unified Germany is strongly on the heels of both Japan and the United States and is expected to overtake both in technology exports by 1995 because of its large technology development momentum.

The problem the United States is facing is that it achieved its greatest technological advances through the efforts of individual inventors like Thomas Edison, Elmer Sperry, and others, who not only researched critical problems but carried their inventions through innovation to application, often as independent researchers and entrepreneurs. They presided over the invention and innovation of the whole technology development from scientific breakthrough to the design and use of the technology as a system. They were not constrained by organizations, rules, or legal restrictions. They worked alone and not as a cog in a large corporate or government organization. They decided themselves what problem to attack and how to use the solution. By and large, they were successful in obtaining the necessary financial support, and some became very wealthy.

Today most researchers and engineers in the United States work as employees of large organizations and are told what to research or engineer. Even the choice of academic researchers is largely driven by the availability of funding or support in particular areas which in turn are decided by outside institutions or organizations. This situation has reduced much of the creative entrepreneurial drive that advanced the United States to its premier technological standing. Strict regulation, interference by legal practitioners, and short-term profit-driven objectives are other reasons for the loss of momentum in U.S. technology development.

The U.S. economy has sustained one of the longest periods of virtual economic stagnation, with an economic growth rate that, for over a decade, has lagged behind the growth of the population. Productivity similarly is barely growing, notwithstanding major technological changes. We no longer value or prefer to buy U.S. products, and much of the pride of everything American is evaporating, at least as far as manufactured products are concerned. At the same time, a decreasing percentage of Americans work in productive sectors of the economy, such as manufacturing and agriculture, while most find employment in the service sectors. The reason for this breakdown is largely found in the lack of competitiveness of U.S. manufacturing, which is caused by a lag in product, process, and management technology.

The industrial sector of the United States has shrunk to less than 28 percent of the total economy and lags behind health care, education, and legal services, which together accounted for over 33 percent of the economy in 1991. At a time when technology is recognized to be the most important factor in economic growth, the United States increasingly emphasizes service sectors that benefit comparatively less from developments in technology. This emphasis not only affects the future of the U.S. economy and thereby the U.S. standard of living but influences our educational and social policies. The result, unless the trend is reversed, may be a long-term decline of the United States as the world's premier economy and technological leader.

13

The Meaning of Technology and Engineering

In this technological age, we face the question of the meaning of technology and engineering, both as a universal and personal inquiry. Since early times humans have attempted to rise above satisfying their basic needs. People in general and those with technical skills often have considered the meaning of technology and the engineering required to make technology happen as a social contract, or an explicit or implicit obligation, to contribute to the betterment of humankind, its way of life, and its environment. Engineers have traditionally shown a hunger for betterment and technical progress which offers opportunities for improvements in the quality of life. Generally, this desire was combined with an urge toward perfection, an elusive objective that engineers thrive for but never achieve because of their own shortcomings, economic pressures, or the lack of patience exhibited by technology users and developers.

As we advance technologically, we increase the danger of confrontation with each other and nature, an upsetting of natural balances and the harmony of our environment. We increase the risk of taking more than we return and thereby depleting the balance of natural resources for generations to come.

As people become more impatient for new technology, they show a declining concern with the harmony of life, of the maintenance of the balance of the world's resources. The rush toward more technology has led to a larger gap between technology use and comprehension as people become increasingly bewildered and confused consumers and users of technology, with little if any idea of what makes it work, what it does, and what effects it may have, both in the short and the long term.

A time of reckoning with technology and its effects is at hand as the ignorance gap between technologists and users of technology widens. Society has developed expectations from technology and a thirst for technological stimuli. Technology for many has become the driving force of progress and of quality of life.

Technology today deals mainly with improving or facilitating what we interpret as the quality of life and not with the problems of the nature of life and what sustains life. The meaning of technology and engineering is to solve the problems of the nature of life, the needs for renewable resources required to sustain life, and the improvements necessary to better the harmony of people and their environment.

The most important aspect of life is change, and the engineer's role is to solve the problems of change. To live fully, people and engineers must contribute to change in a positive, constructive way to assure effective progress for humanity, with an effective balance and harmony of the earth's environment. People who hold back change or do not accept the need for change usually oppose progress. These are often the same people who misuse technology or develop and use it only for short-term gain without respect for the environment or the technology developers.

Engineers must learn to make the right decisions and give technically and morally correct advice and not attempt to simply curry favor or be popular if that advice compromises a technically and morally correct decision. The pursuit of technological excellence and engineering leadership do not conflict. Technological excellence and engineering leadership require hard—often unpopular—decisions, an effective integration of concern with technological objectives, scientific principles, and moral values. Quality is not only a physical and operational characteristic associated with how technology is designed, made, and able to serve; it is a characteristic of engineering decision making. Moral quality implies adherence to right instead of popular decisions, and engineers more than most are required to make courageous, often difficult, decisions, including mistakes on the way toward creative technological solutions.

Technological excellence requires that the technology contribute to the quality of life, human progress, and the maintenance or improvement of the environment. The daunting task is to not only develop but also apply technology to such areas as manufacturing and services. To assure that the time between technology development and application is brief and that the process is continuous, technology must be used more effectively in the leveling of social inequities, the gap between rich and poor, the vast differences in education, health care, and social services.

Technology developed solely for narrow profit objectives often provides solutions to unimportant or nonexistent problems. Technology development objectives must include social goals and environmental considerations, in addition to economic and financial objectives to assure relevant problem solutions and a meaningful contribution of technology to the betterment of human life.

Individualism made this country great but is increasingly disdained. Individualism in engineering and technology development is a positive force only when combined with moral strength and social responsibility. The increasing lack of these fundamental requirements makes individualism in the United States a force for self-destruction, which in recent years has led to inadequate cooperation among engineers themselves; among engineers and marketing, production planning, and management professionals; and among society in general.

14

Threats to U.S. Technological and Economic Leadership

The victory in the Gulf War not only was a result of superior technology but was caused largely by the strength of American ideals, policy, and leadership that forged an unbeatable alliance representing democratic capitalism. The demise of the Soviet empire again was largely the result of U.S. leadership in freeing the world from dictatorship, human oppression, and poverty.

The United States is the logical choice to continue the lead of a more unified world toward improved living standards, equity, and freedom, but the role may be assumed by others if the United States fails to strengthen its economy and resume its technological leadership. With the end of the Cold War, the new world order will be established on the basis of economic advances, and the United States may miss the boat unless it reestablishes its technological leadership which, today, is a requirement for economic leadership.

U.S. productivity is still anemic, budget deficits are rising dangerously, the national debt is well above half the GNP, and trade as well as current account deficits continue to grow dangerously. Technology and productivity are the two engines of growth in wealth and living standards. Without improvements in productivity and renewed technological leadership, Americans may well lose their accustomed way of life and role in the world. Investment and technology are the stimulants of growth and leadership, and they in turn are fostered by education, research, and development. Rivals of the United States spend more on the research and development of new processes than the United States, which devotes 70 percent of its R&D to inventing new products, often failing to bring them to market in a timely and effective manner.

Similarly, investment in meaningful education or real learning is by far greater in Japan, Germany, and other industrialized countries. They not only invest in the basics of education but also instruct their young in discipline and responsibility and instill them with initiative, motivation, and pride as part of their education (see Appendix C).

The threats to American leadership are real and pervasive. The U.S. share of the world economy now stands at 23 percent; without reversing recent trends in investment and productivity, the U.S. economy will decline and with it U.S. economic leadership and its citizens' standard of living. The change in standards of living expressed in real gross domestic product (GDP) per capita (in 1980 dollars) is expected to grow between 1990 and 2000 by 21 percent for the United States, 40 percent for Japan, 27 percent for all of Europe, and 33 percent for Germany. In fact, the United States with $17,252 per capita is expected to lag behind Germany ($18,025 per capita), Japan ($18,055 per capita), and a host of other industrialized countries. This surely is a reversal as the U.S. GDP per capita exceeded that of its nearest rival by over 20 percent in 1982. By the end of the twentieth century, U.S. participation in the world economy is projected to fall by nearly 2 percent to 21.4 percent while Japan and Europe will contribute 10.7 percent and 17.9 percent, respectively (U.S. Department of Commerce, 1991).

Trade balances are expected to continue to favor countries such as Japan and Germany, which now have surpluses of $64 billion and $72 billion, respectively, while the United States will sustain a trade deficit estimated at $109 billion. While overall U.S. productivity is still slightly ahead, Germany, Japan, and others are able to drive up productivity much faster than the United States, using a greater focus on education, teamwork, and technology. U.S. productivity improvements were anemic in the last few years, while major competitors achieved productivity gains of 3.8 to 5.2 percent per year consistently. This gain more than made up for a higher increase in unit labor costs.

Another threat to U.S. productivity is the increased involvement in research and development by competitors of the United States. While the United States led all industrialized nations in the number of scientists and engineers engaged in research and development until the early 1980s with about 0.6 percent of the labor force, Japan and Germany now equal U.S. involvement in R&D at about 0.7 percent of the labor force. Similarly, Japan's and Germany's nondefense funding for R&D is now well over 3 percent of GNP versus the U.S. level of about 2 percent in 1990. Japanese and German research and development is oriented more toward product innovation and application and is focused on manufacturing process development and to a lesser degree on basic science and defense-oriented research (see Appendix F). As a result, the distribution of patents granted has shifted. While the United States granted 67 percent of all the world patents in 1975, its contribution fell to 53 percent in 1989, with Japan increasing its contribution from 8 percent in 1975 to 21 percent in 1989.

Engineering education in particular must be strengthened in the United States. According to the National Science Board (1989), the United States awards a much smaller percentage of undergraduate degrees in science and engineering than are awarded in Germany or Japan—only 58.8 percent as many engineering and science degrees as Germany and 75.0 percent as many as Japan. Even more serious is the content or level of both high school and undergraduate education which increasingly lags far behind that of other major industrialized countries.

One factor appears to be the significantly shorter school year or days spent in school. U.S. schools require 180 days versus 243 in Japan and 190 in Germany. The discrepancy at the university level is even greater.

Engineering science, which has served as the main approach to engineering education during the past 50 years, introduced greater analytic rigidity and discipline to engineering and moved it from a largely empirical to a scientific base. While this shift has provided great new insights into many aspects of engineering and permitted rapid advances in most areas of technology, it has forced a large degree of focus and specialization which often caused engineers to become narrow specialists, without an adequate understanding of the functions of the technology they were contributing to or the environment in which such technology would perform.

As technology and technological systems have grown larger and more complex, increasing problems of systems integration have become evident. To cope with this new development, engineers will have to learn more about technology integration; how to match technology to its environment; how to assure effective design and production coordination; and how to contribute to technology development, implementation, and use in a global environment.

Another issue the United States must consider is labor-management relationships which affect not only workdays lost as a result of disputes but also worker and staff attitude and motivation. U.S. workday losses were about ten times greater than Japan's on a percentage of total annual workdays and over twenty-two times those lost in Germany.

Most serious is the difference in investment in plant and equipment; in the United States 11.2 percent of GDP is so invested, well below Japan with 25.1 percent of GDP in 1990. This gap continues to widen and has long-term implications as it assures that U.S. industries will work with increasingly older processes and technologies. These threats impair U.S. economic and technological leadership, but at the same time they are generating the bugle call for action. The United States is rapidly awakening from its complacency and tolerance. U.S. industry and labor are starting to pull themselves together. Firms are becoming meaner, leaner, and more aggressive. Labor is becoming more responsive. The laggards are now U.S. institutions and bureaucracy, including educational and research institutions which, by and large, have been slow in recognizing the need for change.

The change requires discipline, focus, and a new commitment to quality in education without compromise, efficiency in bureaucracy or administration without fail, and effectiveness in the teaming of labor, management, and government. Adversarial and confrontational relationships, indifference to quality in education and work, lack of teamwork, and ineffective development and use of technology must become things of the past.

The role of the engineer as a leader toward economic growth must be revived. Without promoting and developing, a new role for engineers' technological leadership will be hard to accomplish. But the new engineer will have to be a professional who does not lose sight of the forest in developing the tree.

15

The New Engineer

The engineer of the future will be a well trained, competent, ethical, and assertive professional able and willing to assume responsibility, to lead, and to make decisions. Though an expert in certain areas of engineering and technology, he or she will be well acquainted with the problems technology is to address and the environment technology will face. He or she similarly will be able to determine and trade off the costs and benefits of new technology in the narrow and broad sense, design and introduce technology in a way that makes it difficult or impossible to misuse it, and lead technology policy making at the local or plant as well as the national and global level. Only then can we rest assured that engineering and technology will be developed and used more wisely, safely, and effectively in the interest of humankind.

To achieve technological excellence, engineers will have to return to their original calling and become ingenious and single-minded problem solvers who resolve increasingly complex issues and make decisions only technologically trained people or leaders can make in this technological age. They will have to recognize the broader issues that must be addressed and the environment that must be considered.

Technological excellence will not be achieved if engineers are not willing to assume leadership and its concurrent responsibilities. Engineers will have to consider themselves custodians of the public interest.

The key to economic leadership is technological excellence which in turn depends on the effectiveness of engineers as leaders in a new forward-looking, committed, and energetic society. The challenge is not only how to train such new engineers but how to convert them to their new leadership role. Concurrently, society must be convinced that to reap the fruits of advancing technology it must vest greater responsibility and trust in technologists and engineers—otherwise the fruits of technology may become a bitter harvest.

Engineering leaders are those who develop effective ways to use technological products and processes and persuade others to choose and use technology wisely. Leaders require maturity, communication skills, compassion, a willingness to understand others and their viewpoints, as well as professional competence.

Engineering leaders must be able to transfer or share credit, admit mistakes, and make unpopular decisions when necessary. They must be able to consider the broader implications of a decision while still conversant with the actual engineering requirements for its implementation.

Engineers in the recent past have not been effective leaders, largely because of their concern with or concentration on narrow engineering and scientific issues. This focus must change if we are to regain technological preeminence which is only possible with greater engineering leadership in government and industry.

Engineers must be willing to take an interest in the broader issues affected by or effecting technology which, as noted, does not require a dilution of professional engineering competence but simply greater maturity. Engineers must be trained to and willing to take responsibility, make decisions, and guide policy. To accomplish this, the education of engineers must change as must the role engineers play in organizations, industry, and government.

We have made some progress in training nontechnical leaders in industry and government in the management of technology and in the role of technology in society, but ultimately we need professional engineers with both a fundamental knowledge of engineering principles and an understanding of the capabilities, constraints, impacts, and potential developments of technology to make the really hard decisions that more and more are introduced by changes in technology. Engineering leadership, economic development, and technological excellence are interdependent and must be considered so if the real rewards of technological developments are to be reaped.

Appendixes

APPENDIX A
SAFETY AND ENGINEERS' RISK AVERSIVENESS

Nearly everyone is concerned with safety and voices opinions on how to improve it. Engineering societies publish codes of ethics that are concerned largely with safety, while they and others also publish standards of engineering design with an emphasis on design for safety. Some now insist that engineers ought to design 100 percent safety into anything that will be used by or impacts on society.

The degree of safety designed into a technology is a difficult issue because while one can design a nearly perfect, high-quality, and safe structure or system at no additional cost, designing it for perfect safety is not only very costly but usually impossible; there always are exceptional conditions that cannot be identified during the design or that are so unlikely that it makes little sense to design for them. We have become used to living with such risks in everyday life, including, for example, transportation and health care.

Engineers are very much concerned with safety and will attempt to assure it in their designs. But they are risk averse by training and consequently in attitude which often interferes in the attempt to design rational safety into systems. Overdesign may safeguard a structure but may make it more unwieldy or difficult to use. A heavier than necessary automobile may be safer in a collision but may as a result of its longer stopping distance and greater inertia actually increase the probability of a collision. Safety considerations also may conflict with operability or maintainability of a system.

Engineers usually are not familiar with the often conflicting requirements of safety, reliability, economy, and operability. Where such conflicts exist, safety often will attain a low priority as it is the most difficult to quantify, particularly in monetary or value terms.

APPENDIX B
CODES OF ETHICS FOR ENGINEERS AND THE
ROLE OF QUALITY MANAGEMENT IN
ENGINEERING

Various engineering organizations have developed codes of ethics for engineers. These usually include rules of behavior among engineers and between engineers and their clients. They similarly define responsibilities of engineers toward their work, their coworkers, their employers, and their customers. Many codes define safety or environmental impact standards and usually confine themselves to narrow client, intraprofessional, and similar relationships.

The time has come for us to reconsider engineers' codes of ethics to include all aspects of technological impacts. One particular area is the management of quality in engineering.

Quality concepts are inherent in the requirements for engineering and include not only on-time delivery of engineering output but also consideration of the level of accuracy expected. Preparing for total quality management (TQM) engineering demands a more detailed specification of engineering activities and their feasible on-time performance, in line with the processes they define or serve. Organizing for TQM in engineering requires complete integration of all engineering activities into the overall activities of engineered systems.

TQM techniques are designed to permit standards and requirements to be applied uniformly and fed both back into engineering and forward from engineering into manufacturing and use. Implementing TQM requires the use of TQM techniques to reduce waste, to assure effective coordination during engineering, and to permit just-in-time delivery of the intermediate and end results of engineering.

APPENDIX C
A NATION NEITHER PREPARED NOR COMMITTED
TO THE TRAINING OF ITS WORKERS

Inevitably, U.S. industry is moving toward higher technology, greater use of automation, and improved efficiency, with a resulting loss of low-level manufacturing and middle-level management jobs. These are replaced by a smaller number of jobs requiring modern higher technological skills. In the 1980s the United States lost over 1.9 million manufacturing jobs, plus 0.2 million middle-management industrial positions. At the same time fewer younger workers joined the manufacturing industries just when younger, more educated workers were required. The number of workers 24 years or younger fell from 18 percent in 1981 to just under 11 percent in 1991. Similarly, the percentage of workers under 35 years of age is declining precipitously in the manufacturing industries. A related issue is the low level of women and minorities such as blacks and Hispanics among manufacturing workers.

The average age of the American workforce will rise to 40 in 2000 from 28 in 1970. Similarly, the average age of employed engineers rose to 44 years in 1990, from only 32 years in 1970.

We now have an increasingly older, less educated, largely male, and white workforce. Even though the average age of manufacturing workers and lower/middle-level managers is now well above 45 years of age, many employers do not invest in training of older workers. As a result, few older workers or the bulk of the U.S. manufacturing workforce are offered no informal opportunities to improve their skills, according to the U.S. Office of Technology Assessment's 1990 report on worker training. Where training is offered, it is often without commitment and on an impromptu basis. Furthermore, it usually is provided by outsiders with little experience of real shop floor problems and, therefore, it has little applicability to issues faced by the workers. In addition, training is seldom structured to orient workers toward particular work or career opportunities. The relationship among training, job performance, and position is generally kept vague. As a result, workers have become skeptical toward training.

U.S. industry needs people capable of handling jobs in the growing complexity of knowledge-intensive manufacturing. While retraining of the existing workforce may yield the best short-term rewards, the average worker is over 45 and may be unable or unwilling to absorb new knowledge.

APPENDIX D
ON CONTINGENT FEES AND SELF-REGULATION[1]

We acknowledge that the practice of law is a monopoly in a sense that only those licensed may lawfully practice law. We know, too, that it is an article of faith in our country that monopolies must be regulated.

But regulation of the practice of law, like that of medicine, and of some other professions, has been left largely to the professions—up to now. Of course, a profession has an obligation to regulate itself. Regulation from the outside has come about only when there was overstepping of the bounds, and when the public interest required action which the professions themselves failed to take. . . .

Few things have done more serious damage to the standing of the legal profession than the unseemly—indeed shocking—spectacle of open solicitation by a handful of lawyers who dashed off to India to solicit clients after the tragic multiple disaster in Bhopal. . . .

Another subject also in need of careful scrutiny is contingent fees. So far as I have been able to discover, we are the only country where that practice is accepted. I have said before, and I repeat now, that the first duty of a lawyer in dealing with a client, whether in relation to a personal injury or other claim, is to inform the client of the probability of success as nearly as that can reasonably and fairly be evaluated.

Every person in this room knows that in certain kinds of multiple disaster cases there is likely to be little question about liability. In such cases there is no place for contingent fees of the kind now widely practiced.

Many honorable lawyers have recognized this by agreeing to charge a fee not to exceed a fixed percent, with the final charge based upon the time involved and results obtained. I do not, therefore, suggest that contingent fees be outlawed or abolished, but I do suggest that any study . . . should consider whether contingent fee agreements should be made subject to the control of the court having jurisdiction of the claim. . . .

There has also been criticism of the bar for the lack of adequate disciplinary proceedings. Up to this date, we have not done so adequately.

I am well aware that the need for stronger disciplinary measures is by no means limited to practitioners. In recent years, members of the Judiciary, both state and federal, have been found guilty of violating criminal statutes.

In cases involving federal judges, the Judicial Councils of the Circuits have moved promptly to deal with this problem. The misconduct of a few judges–state and federal–has helped to bring about the present state of public reaction toward the profession as a whole.

I sense that the hostility toward our profession is growing and has grown sharply in the last eight or 10 years. One of the consequences is an increasing demand for legislation to regulate the profession. For nearly 200 years legislatures have been content to allow us to regulate our profession. In 1969, the [American Bar] Association, not some legislative body, took the initiative in the revision of the Code of Judicial Ethics, and the more recent model rules of professional Conduct. We must continue this pattern of responsible self-regulation.

NOTE

1. This appendix is excerpted from remarks to the 1986 annual meeting of the American Bar Association on August 11 by retiring U.S. Supreme Court Chief Justice Warren E. Burger.

APPENDIX E
ENGINEERS' LIABILITY

Engineers that are personally responsible for actions that cause harm to others may be liable to third parties for injuries that result from a design defect. Such liability usually includes claims for property damage. This liability can be applied equally to engineers working for themselves, for a design firm, or as employees of a corporation. In other words, employment status by a corporation will not by itself insulate an engineer from liability. As a result, engineers may have to seek personal insurance coverage against professional claims even if employed by a corporate entity unless the corporation explicitly provides such insurance.

APPENDIX F: COMMERCIALLY RELEVANT R&D

In recent years, the United States has fallen well behind other industrialized nations in its support for the development of commercially relevant technologies.

United States	1.8 percent of GNP
Japan	3.0 percent of GNP
Germany	2.8 percent of GNP

More important, while U.S. expenditures for commercially relevant R&D are falling, those of Japan and Germany continue to rise. Furthermore, the amount spent on product and process innovation and engineering in Japan and Germany is about double that spent in the United States as a percentage of GNP.

Bibliography

Aldridge, M. D. "Technology Management: Fundamental Issues for Engineering Education." *Journal of Engineering and Technology Management* 6 (1990).

Allen, T. J. *Managing the Flow of Technology.* Cambridge, Mass.: MIT Press, 1977.

Anderson, A. *Technical and Technological Education in Japan.* Japanese National Commission for UNESCO, December 1972.

Bennis, Warren. *The Unconscious Conspiracy.* San Francisco: Jossey-Bass, 1976.

———. "Leadership Transforms Vision into Action." *Industry Week*, May 31, 1982, 54–56.

———. *Why Leaders Can't Lead: The Unconscious Conspiracy Continues.* San Francisco: Jossey-Bass, 1989.

Betz, F. *Managing Technology.* Englewood-Cliffs, N.J.: Prentice-Hall, 1987.

Bhalla, K. B. *The Effective Management of Technology.* Reading, Mass.: Addison-Wesley, 1987.

Dietrich, W. S. *In the Shadow of the Rising Sun: The Political Roots of American Economic Decline.* Philadelphia: Pennsylvania State University Press, 1991.

Dorman, A. A. "From Here to Uncertainty—Do You Really Want to Be an Engineering Manager?" *Journal of Management in Engineering* 4, no. 4 (October 1988).

Frankel, E. G. *Management of Technological Change.* Boston: Kluwer Academic Publishers, 1990.

Ginn, M. E. "Creativity Management: Systems and Contingencies from a Literature Review." *Engineering Management* EM-33, no. 2 (1986).

Grayson, L. P. "Leadership or Stagnation? A Role for Technology in Mathematics, Science and Engineering Education." *Engineering Education* 73, no. 5 (February 1983): 356–66.

Hazelrigg, G. A. "In Continuing Search of the Engineering Method." *Engineering Education* 78 (1988).

Hazen, H. L. "America Meets Japan in Engineering Education." *The Technology Review* (May 1952): 351.

———. "The 1951 ASEE Engineering Education Mission to Japan." *Journal of Engineering Education* (June 1952).

Hilton, R., and Lee, I. Office of Technology Assessment, U.S. Congress. "Educating Scientists and Engineers." OTA Report Brief, Superintendent of Documents. Washington, D.C.: U.S. Government Printing Office, June 1988.

Kanter, A., and Mirvis, S. *Cynical Americans: Living and Working in an Age of Discontent and Disillusion.* New York: Jossey-Bass, 1990.

Kemper, J. D. *Engineers and Their Profession.* Orlando, Fla.: Holt, Rinehart and Winston, Saunders College Publishing, 1990.

Koen, B. V. "Definition of the Engineering Method." *Engineering Education* (1985).

Morrison, E. E. "Triumphs and Shortcomings of Engineering." *Technology Review* (June 1982).

National Academy of Engineering (NAE). *Public Attitudes Toward Engineering and Technology.* NAE Office of External Affairs, 202–334–2210 (1986).

National Science Board (NSB). *Science and Engineering Education in the U.S.* Washington, D.C.: National Science Board, 1989.

National Science Foundation. *Technological Innovation in the U.S.* Washington, D.C.: National Science Foundation, 1976.

National Science Foundation and the U.S. Department of Education. *Science & Engineering Education for the 1980's & Beyond.* Washington, D.C.: National Science Foundation, 1980.

Perruci, R., and Gerstl, J. E., eds. *The Engineers and the Social System.* New York: Wiley, 1969.

Smith, O. E. "The Need for a National Technology Policy." *Research Manager* (June 1986).

Susskind, C. *Understanding Technology.* Baltimore, Md.: Johns Hopkins University Press, 1973.

Tribus, J. "Deming's Way." *Mechanical Engineering* (January 1988)

U.S. Department of Commerce, Office of Industry and Trade Information. *U.S. Foreign Trade with Leading Countries.* Washington, D.C.: U.S. Department of Commerce, 1991.

Wills, C. D. "How Engineers Can Learn to be Better Managers." *APWA Reporter,* October 17, 1981.

Index

ern industrialized countries, 8, 69–70. *See also* Engineering education
Education Amendments of 1972, 74–75
Employment, technological change and, 19, 162
Engineering: creativity and, 47–51, 57, 103–4, 113; diverse fields in, 91; effective organization of, 56–57; evolution of, 62–63; excellence in, 118–20, 126; management interdependent with, 41; marketing of, 46–47; MIT seminar and, 9; public attitudes toward, 35; qualifications to practice, 91, 92–93; science and, 77; status of, East-West comparisons and, 72; technology development role of, 61; total quality management of, 116, 161; traditional concept of, 13–14; U.S. Congress and, 5; U.S. education and, 26; U.S. military and, 3; U.S. product competitiveness and, 64; Western education and, 8
Engineering associations, 57–59, 91–92, 96–97
Engineering education: changes needed in, 83–86; creativity and, 68; curriculum for, 16–17, 88–90; decision making and, 68; engineering leadership and, 68, 84–85, 156; engineering management and,

117–18; engineering objectives and, 42; in Europe, 67, 72, 112; evolution of, 68; focus of, 68–69, 83, 113; in Germany, 114; industrial experience and, 78; industry-funded, 112; in Japan, 71–72, 75, 82, 105, 111–12, 114; knowledge upgrading and, 116; military and, 85; problem solving and, 80; science and, 68; societal needs and, 68; technology development and, 68, 90, 117; technology use and, 84; timeliness and, 117; Western industry and, 85. *See also* U.S. engineering education
Engineering leadership: aversion to change and, 16–17; characteristics of, 31–34, 156–57; as change requirement, 153; communication skills and, 41, 45–46, 96; criticism and, 31; economic growth and, 138, 140, 157; engineering education and, 39, 84–85, 156; engineering goals and, 42; engineers' focus and, 156; engineers' roles and, 14–15; ethics and, 38; experience and, 33; interpersonal skills and, 45–46, 96; Japan and, 15; lack of, 122–23; leadership skills and, 30–31, 34, 41–42; management skills and, 41; morality and, 146, 147; necessity of,

2, 11; effectiveness of, 3; employment and, 19; engineers'
role in, 22–23; environmental
impact of, 22, 145–46; evolution of, 2; Far Eastern countries and, 8; industrial growth
and, 1, 11; management of, 1–
2, 11; population explosion
and, 21; rejection of, 21–22;
risk of, 2; social structure
and, 12; societal impacts of,
1, 22–23, 121. *See also* Technology development
Technological excellence: constituents of, 135–36; creativity and, 136; economic
growth and, 10, 140, 157; engineering leadership and,
136–37, 138, 140, 146–47,
155–56, 157; management of
technology and, 6; necessity
of, 104; people management
and, 133–34; planning for,
106–7; quality of life and,
147; requirements for, 140;
teamwork and, 136; U.S., requirements for, 109
Technological innovation, 7
Technological invention/discov-
ery, 12–13, 18–19
Technology: Americans' faith
in, 35; complexity and, 131;
decision makers' conversance
with, 22; definition of, 62;
economic growth and, 22,
149–50; engineering education and, 68, 83, 90; environmental impacts of, 36,

120–21, 122; Gulf War and,
149; irrelevant, 121; knowledge and, 135; management
of, 5, 6, 11, 19 (*see also* Engineering leadership); meaning
of, 145, 146; misuse of, 83,
120–22, 146; obsolescence
and, 104; public attitudes toward, 34–35; quality of life
and, 146; robotics, 107, 114;
scientific knowledge base of,
61; societal impacts of, 121;
software and, 135; U.S. Congress and, 5; world trade and,
22
Technology development: automobile industry and, 133; engineering education and, 68,
90, 117; engineering research
and, 81; engineering's importance to, 117–18; engineers'
changing role and, 62–63;
engineers' effectiveness in,
61–62; engineers' integrative
function and, 62–64; environmental considerations and,
147; impacts of, 100, 120–
21; individualism and, 147;
industrial leadership in, 18–
19; Japan and, 105–6, 110,
113; liability claims and, 94;
management and, 84; objectives of, 147; planning and,
107; problems caused by,
120–21; social goals and,
147; teamwork and, 128,
136; timeliness and, 114,
128, 133, 136; U.S. leader-

About the Author

ERNST G. FRANKEL is Professor of Ocean Systems at the Massachusetts Institute of Technology. He is the author of six books including, most recently, *Management of Technological Change* (1989).